PIPELINE DREAMS

PEOPLE, ENVIRONMENT, AND
THE ARCTIC ENERGY FRONTIER

Mark Nuttall

IWGIA – Document 126

Copenhagen – 2010

PIPELINE DREAMS
PEOPLE, ENVIRONMENT, AND THE ARCTIC ENERGY FRONTIER

Author: Mark Nuttall
Editorial Production: Kathrin Wessendorf and Cæcilie Mikkelsen
Copyright: © Mark Nuttall and IWGIA
    2010 – All Rights Reserved
Cover and layout: Jorge Monrás
Cover Photo: Trans-Alaska Pipeline by Travis Shinabarger
Maps: Jorge Monrás
Proofreading: Elaine Bolton
Prepress and Print: Eks-skolens Trykkeri, Copenhagen, Denmark
ISBN: 978-87-91563-86-7
ISSN: 0105-4503

Distribution in North America:
Transaction Publishers
390 Campus Drive / Somerset, New Jersey 08873
www.transactionpub.com

Hurridocs cip data

Title: Pipeline Dreams: people, environment, and the Arctic energy frontier
Author: Mark Nuttall
Corporate Editor: IWGIA
Place of publication: Copenhagen, Denmark
Publisher: IWGIA
Distributors:
North America: Transaction Publishers – www.transactionpub.com
UK: Central Books Ltd. - www.centralbooks.com
Books also available from IWGIA - shop.iwgia.org
Date of publication: December 2010
Pages: 223 – maps
Reference to series: IWGIA Documents Series, no. 126
ISBN: 978-87-91563-86-7
ISSN: 0105-4503
Language: English
Bibliography: yes
Index terms: Indigenous peoples/ Hydrocarbon extraction/ Environment/ Sustainable development/Indigenous rights
Geographical area: Northern America
Geographical code: 0204

**INTERNATIONAL WORK GROUP
FOR INDIGENOUS AFFAIRS**
Classensgade 11 E, DK 2100 - Copenhagen, Denmark
Tel: (45) 35 27 05 00 - Fax: (45) 35 27 05 07
E-mail: iwgia@iwgia.org - Web: www.iwgia.org

IWGIA

PIPELINE DREAMS

Contents

CHAPTER I

INTRODUCTION

Dreams of extracting oil, gas and minerals and developing the resource potential of the circumpolar North have been significant in shaping relations between indigenous and non-indigenous peoples, as well as in the opening up of Arctic frontier regions in Canada, Alaska, Siberia and other northern areas to economic development. Yet oil and gas development is only one event in a chain of historical events that have transformed indigenous societies and communities throughout the world's Arctic and sub-Arctic regions. In northern North America, understanding the history of the fur trade, of settlement by non-indigenous people, of pushing back the frontier, of the Gold Rush, of searching for minerals and fossil fuels, or the construction of military infrastructure during the Cold War—in Alaska, Yukon and the Northwest Territories (NWT) as well as in northern Alberta and northern British Columbia—is extremely relevant to contemporary discussion of oil and gas exploration and energy development because the experience of past development and the impact of megaprojects tells us a considerable amount about the sustainability of industrial economies, the impacts of resource exploitation in fragile ecosystems, and the profound consequences of environmental, social and economic change in northern communities (Piper 2009).

Indigenous peoples have long been involved in struggles to make sense of, adapt to, and negotiate the impacts and consequences of resource development and the extractive industries, but have also been involved in struggles to gain some measure of control over development as well as to benefit from it.[1] Flyvbjerg et al. (2003: 5) argue that "project promoters often avoid and violate established practices of good governance, transparency and participation in political and administrative decision making, either out of ignorance or because they see such practices as counterproductive to getting projects started. Civil society does not have the same say in this arena of public life as it does in others; citizens are typically kept at a substantial distance from megaproject decision making. In some countries this state of affairs may be slowly changing, but so far megaprojects often come draped in a politics of mistrust." Today, in the face of intensifying environmental and economic changes, the local observations, perspectives and concerns of northern indigenous peoples regarding oil and gas development activities and their consequences are crucial to bring to the dialogue about the place and importance of the Arctic and sub-Arctic for the global energy future. In a contribution to this dialogue, this book includes consideration and discussion of both positive and negative appraisals of social and cultural impacts of oil and gas development, together with community concerns about the effects of this on the environment and wildlife and indigenous people's daily lives. Although the focus is largely on northern Canada, it also discusses cases from Alaska and elsewhere in the circumpolar North.

Arctic Oil and Gas

Significant changes in world energy markets, in increasing global demand, and advances in oil and gas industry technology, have led to a major expansion of oil and gas exploration and development in many parts of the Arctic over the last thirty years. This activity looks set to intensify, especially as some energy analysts suggest that the world's existing oil reserves may well not be enough to meet demand over the next 15 to 20 years. According to recent estimates by the U.S. Geological Survey, the Arctic may contain 25% or more of the world's undiscovered natural gas reserves and 13% of its untapped oil. This means that the Arctic's undiscovered conventional oil and gas resources are estimated to be approximately 90 billion barrels of oil, 1,669 trillion cubic feet of natural gas, and 44 billion barrels of natural gas liquids.[2] The energy industry seems increasingly prepared and technically-equipped to meet the challenges of operating in the Arctic's harsh and demanding terrestrial and marine environments (although the Deepwater Horizon accident in the Gulf of Mexico has caused many to question this) and the region is being imagined and defined as the next energy frontier. The circumpolar North becomes even more attractive to energy companies just as a combination of factors—depletion of existing reserves in places such as the North Sea (as well as concerns that global oil production will peak in the near future), local conflicts in places such as the Niger Delta, and geopolitical tensions in the Middle East being just a few— make it more difficult for industry to continue to invest and work in areas which have, until now, provided much of the world's oil and gas.

Resource scarcity often has nothing to do with the physical shortage of oil and gas but with ecological and economic limits, as well as political issues, such as higher exploration and drilling costs, the rising costs of moving supplies, and the local and regional difficulties of production, which all combine to make returns problematic (Pratt 2001).[3] Political talk of a looming global energy crisis highlights the fact that energy security and energy independence are also concerns for many countries that all too often feel vulnerable because of their reliance on oil and gas from other places. On New Year's Day 2006, for instance, Russia cut off gas deliveries to Ukraine, the main conduit to western Europe, prompting fears of a winter energy crisis in Europe. Western European media called it a "Cold War-style threat". Although the energy crisis did not happen—Europe's gas supply was turned on a couple of days later—the situation nonetheless highlighted fears of insecurity in Russia's energy sector and the uncertainty of supply, as well as concerns over the control of oil and gas resources to exert political pressure. In his assessment of America's dependency on imported oil, Michael Klare writes that: "It doesn't take a vivid imagination to grasp the essence of America's energy predicament: only the Middle East and other regions that have long suffered from instability and civil unrest have sufficient untapped reserves to satisfy our (and the world's) rising petroleum demand in the years ahead. Like it or not, for as long as we continue to rely on petroleum as a major source of energy, our security and our economic well-being will be tied to social and political developments in these unpredictable and often unfriendly producers" (Klare 2004: 20).

American strategic interests in the Middle East and Central Asia, as well as moves by other countries to secure access to—and ultimately to control—oil and gas supplies, have prompted commentators to write and speculate about the future of energy reserves in terms of resource wars and a new Great Game (e.g. Klare 2004, Kleveman 2003, McQuade 2004). Writing about the politics of energy in Central Asia, Lutz Kleveman says that "great empires once again position themselves to control the heart of the Eurasian landmass, left in a post-Soviet power vacuum. Today there are different actors and the rules of the neocolonial game are far more complex than those of a century ago: The United States has taken over the leading role from the British. Along with the ever-present Russians, new regional powers such as China, Iran, Turkey, and Pakistan have entered the arena, and transnational corporations (whose budgets far exceed those of many Central Asian countries) are also pursuing their own interests and strategies" (Kleveman ibid.: 3).

As Arctic sea ice continues to melt under conditions of climate change, and as countries appear to be excited by the prospect of the discovery and development of new energy resources in the circumpolar North, the Great Game has also supposedly moved north. A new Great Game is often spoken about as being played out in the Arctic in one possible scenario for describing international relations in the coming decades. The term is used in an attempt to capture the feeling that countries are jousting for control of resources and that there is a new Arctic "gold rush" as states scramble to claim ownership of the Far North in advance of an irreversible meltdown. It may have more resonance when used as journalistic rhetoric than when describing current or future international affairs in the circumpolar North—the 19th century Great Game involved the British Empire, France and Tsarist Russia and the term more accurately referred to espionage in defence of empire, the incitement of rebellion to destabilize empire, and the control of Afghanistan and the Hindu Kush—but, considering the strategic importance of the Arctic and the increasing interest in the region's resources, not just by Arctic states but by countries such as China and India, as well as European Union member countries, the point is taken.

Northern countries are positioning themselves to assert Arctic sovereignty and lay claim over the Arctic. In recent years, the world's media have reported on an international rush to claim the Arctic Ocean and its surrounding waters—indeed the North Pole itself has become subject to geopolitical posturing. Amid talk of "global energy hunger" and an oil and gas boom in the Arctic, an international "cold war" is said to have begun, characterized by dispute over who actually owns the Far North.[4] Russia and Norway, two of the world's largest net oil exporters, are eyeing largely untapped reserves in the Barents Sea in areas where they have competing claims over sovereignty, although the two countries have recently reached agreement over a decades-long boundary issue. The press, particularly in Canada, picks up every so often on stories about the tension that simmers between Denmark and Canada over Hans Island, a wedge of rock situated between north-west Greenland and the eastern coast of Canada's Ellesmere Island, as well as sovereignty disputes over other parts of the Arctic, such as the Northwest Passage.

Such reports seem to suggest that the assertion of territorial claims and moves to stake claims to rights over lucrative resources have intensified since a team in a subsea craft from a Russian expedition planted the Russian flag on the Lomonosov Ridge in summer 2007. The Lomonosov Ridge extends for some 1,800 km across the entire bed of the Arctic Ocean's central basin, from the region off Russia's Novosibirsk Islands to near the northern tip of Ellesmere Island. Establishing whether the Eurasian and North American landmasses are connected to the Lomonosov Ridge has preoccupied agencies undertaking geological surveys in several Arctic countries. The five countries bordering the Arctic Ocean (the United States, Canada, Russia, Norway and Denmark/Greenland) have been reported to be carrying out geological research so as to establish—and lay claim to—the continental shelves extending from their continental margins. Russia claims the Lomonosov Ridge, but the other states dispute this while similarly laying claim to it and to other parts of the Arctic Ocean.

A simplified argument is that climate change is driving the urgency to make these claims and that, as sea ice diminishes due to global warming, the Arctic Ocean will be increasingly accessible. True, it may well be that reduction in perennial sea ice cover opens up Arctic waters to shipping and increased resource exploration, but the reasons why the five Arctic coastal states have been mapping the extent of their continental shelves have more to do with their obligations under the United Nations Convention on the Law of the Sea (UNCLOS).[5]

Social and Environmental Impacts of Energy Development

The Arctic has been inhabited by indigenous peoples for millennia. They include the Inupiat, Yup'iit, Alutiit, Aleuts and Athabascans of Alaska; the Inuit, Inuvialuit, Athabaskans, Dene and Cree of northern Canada; the Kalaallit and Inughuit of Greenland; the Sami of northern Fennoscandia and Russia's Kola Peninsula; and the Chukchi, Even, Evenk, Nenets, Nivkhi, Yukaghir and many other groups of the Russian Far North and Siberia. Arctic peoples have depended for thousands of years on the living resources of land and sea, as hunters, fishers and reindeer herders, and today customary resource use practices remain of crucial importance for local economies and cultures (Nuttall 1998, Nuttall et al. 2005).[6]

The future development of Arctic resources alarms indigenous communities, conservationists and environmental groups already preoccupied with lobbying northern states to protect the Arctic and its wildlife from contaminants and the impacts of climate change. The Arctic has an environmental history of sensitivity and vulnerability to change and to the impacts of industrial development. It is a place with a fragile ecology where environmental scars from resource extraction take decades to heal. In addition to the direct and immediate impacts on the ecology of the Arctic—specifically to vegetation and wildlife habitat—oil and gas development activities have many cumulative effects on traditional resource-use practices and on the economies and well-being of indigenous and local peoples.

More and more indigenous communities, for whom the North is a homeland rather than a resource frontier, are engaged, or are attempting to engage, in dialogue with one another and with government and industry and seek to express their views about what energy development could mean for both present and future generations in terms of socio-economic impacts, community sustainability, wildlife, and environmental health. Some of their concerns about energy development arise from fears of drastic and long-lasting social, economic and environmental impacts, but other concerns are expressed because of disputes about the ownership, use, management and conservation of traditional lands and resources in the homelands of indigenous peoples. These issues are at the heart of the Inuit Circumpolar Council's (ICC) "Circumpolar Inuit Declaration on Arctic Sovereignty",[7] but they are also emphasized in other statements by indigenous peoples who reiterate the need for industrial developers to recognize their responsibility to indigenous and local communities and to the environment.

Even before the massive oil spill in the Gulf of Mexico, few would have denied that the energy industry causes massive environmental damage and has huge social and economic impacts, especially on the communities and regions where extraction takes place. As petroleum and military development spread throughout the circumpolar Arctic in the latter half of the 20th century, transportation infrastructure (such as roads, pipelines, airstrips and ports) contributed significantly to surface disturbance and habitat fragmentation. Chapin et al. (2005) show that between 1900 and 1950, less than 5% of the Arctic was affected by infrastructure development (see also Nellemann et al. 2001). By 2050, somewhere between 50-80% of the Arctic is projected to be disturbed, although this level of disturbance may occur by 2020 in Fennoscandia and some areas of Russia. Furthermore, the energy industry is linked to global climate change—fossil fuels, of course, release not only energy but burning them results in the production of carbon dioxide as well.

As the Arctic continues to be seen as a frontier region for oil and gas development, massive infrastructure will need to be built in areas of ecological sensitivity. The energy industry already moves enormous amounts of equipment into areas that first require transformation into exploration and development sites. Just building these sites has direct impacts such as habitat destruction, the disturbance of animal migration routes and the erosion of landscapes. Freshwater resources have to be drained for the construction of ice roads while gravel has to be quarried and mined to supply material for well pads, roads and harbours. Fears of an environmental disaster have intensified since the Deepwater Horizon accident in the Gulf of Mexico in April 2010. Following an explosion on British Petroleum's Deepwater Horizon drilling rig, which killed 11 oil workers, oil gushed from a sea bed well-head and led to America's worst offshore oil spill. Canada's National Energy Board (NEB), which regulates offshore drilling in northern Canada, has acknowledged that a similar accident happening in the Beaufort Sea cannot be ruled out. Presenting testimony to Canada's House of Commons natural resources committee in May 2010, the head of BP's Canadian operations confirmed that BP did not have a plan in place, nor the capability of preventing or dealing with such an accident in the Arctic (Mayeda 2010). On 11 May 2010, the NEB announced that it would conduct a review of safety and environmental offshore drilling requirements in the Canadian Arctic. The review aims to gather information and knowledge from Aboriginal organizations, residents of Arctic communities, technical experts, governments, regulators and industry, and other participants.[8]

Human impacts and environmental transformation have intensified in the last few decades, intruding on traditional indigenous ways of life and human-environment interactions. Significant climate change is becoming more evident, as is the destructive impact of industry (ACIA ibid.). In Russia, for example, the oil and gas industries are the biggest sources of pollution, affecting reindeer pasture and marine and freshwater environments. Climate change scenarios suggest that the almost complete elimination of multi-year ice in the Arctic Ocean is likely to be immensely disruptive to ice-dependent micro-organisms, which will lack a permanent habitat. It is anticipated that marine mammals such as walrus, seals, whales and polar bears are likely to undergo shifts in range and abundance as the amount of sea ice decreases (Nuttall et al. ibid.). Such changes could have long lasting impacts on the more traditional and customary hunting and fishing economies of many small, remote Arctic settlements. Although the effects of rising temperatures may lead to an increase in biological production in some wildlife species, the distribution and movement of many species that remain crucial to the livelihoods and well-being of indigenous peoples could change. Important wetlands may disappear, or drainage patterns and tundra landscapes may be altered significantly, which could affect waterfowl. Changes in terrestrial vegetation will have consequences for reindeer herding and subsistence lifestyles. Terrestrial wildlife such as caribou and reindeer, two major species important for indigenous communities throughout the Arctic, would be affected by climate change directly through changes in thermal stress in animals, and indirectly by significant difficulties in gaining access to food and water (ACIA ibid., Nuttall et al. ibid.). Arctic communities located on coastlines may be affected by rising sea levels, increased coastal erosion and severe storms.

Although environmental threats to the Arctic associated with oil and gas development, production and transport are primarily local and/or regional rather than circumpolar in scale and extent, important exceptions can occur for certain species of migratory animals if they congregate within relatively small areas affected by intense disturbances (e.g. large oil spills). In such cases, devastating impacts could occur at the population level (Chapin et al. ibid.). Onshore oil and gas activities, such as construction of pipelines and the actual production of oil and gas, also impede access to traditional hunting and herding areas, which disrupts community activities and traditional practices (Brody 1991, Golovnev and Osherenko 1999). Pipelines and their associated facilities, such as gas compressor stations and access roads, create obstacles to the movement and migration of caribou and reindeer herds and impact on traditional harvesting practices. Compressor units may either be field compressors which pump gas from the well-head to a gathering pipeline, or transmission compressors which work to move gas along the pipeline from one compressor station to the next. They generate a loud continuous noise which can disturb wildlife, diverting animals from migration routes and away from traditional hunting areas.

In Russia's North, destruction of vegetation due to facilities, road and pipeline construction, and off-road vehicle traffic in the intensively developed Yamal Peninsula in Western Siberia exceeds 2,500 km^2 and could more than double under current development plans (Forbes 1999). Geographically located in the far north of the West Siberian Plains, in the lower basin of the Ob' River, the Yamal-Nenets Autonomous Okrug (YNAO) was established in December 1930 as part of the Tyumen region. The YNAO covers an area of some 750,000 km^2, including islands in the Kara Sea, with a population approaching 500,000 (around 80,000 live in rural areas, the rest in urban centres). Located in the northern part of the West Siberian oil province, the YNAO has the largest confirmed reserves of oil and gas in Russia. One hundred and thirty-three registered offshore and onshore oilfields constitute about 14.5% of Russia's overall oil reserves. Currently, 33 of these fields are in production and provide just under 9% of Russia's total oil. Natural gas plays a major role in the YNAO, with some 190 registered gas fields in the region. The largest gas condensate fields include Medvezh'e (the first field to be developed, in 1971), Urengoy, Leningradskoe, Rusanovskoe and Purovskoe. This large-scale development has led to an influx of people from southern parts of Russia, resulting in the indigenous population accounting for less than 7% of the overall population (Stammler and Forbes 2006).

Yamal, like many other parts of Russia's North, is an area where reindeer constitute a biological resource of great importance to the physical, economic and cultural survival of indigenous people—indeed, the YNAO is numerically the world's most productive reindeer herding area (Stammler and Forbes ibid.). Around 13,000 mainly Nenets but also Khanty and Komi families herd some 556,000 reindeer in the okrug. The resulting concentration of reindeer herds into an ever-decreasing undeveloped area has led to overgrazing, with potential long-term adverse effects on ecosystem productivity and local economies (Forbes ibid.). Pipeline construction, which creates the need for roads and thereby leads to easier access to formerly isolated regions, also opens up larger areas for additional resource development. Forbes (ibid.) argues that the dual impacts of intensive grazing and industrial development have combined to create a situation whereby the disturbance to the environment is not found anywhere else in the tundra ecoregion. Despite the historic and contemporary situations of coexistence, he argues that energy development and reindeer herding appear to be mutually incompatible.

Research has already shown that oil development has contributed to trends in the changes and behaviour of caribou and reindeer populations in some areas. Caribou and reindeer are sensitive to disturbance during calving (Vistnes and Nellemann 2001; Griffith et al. 2002). In Alaska, for example, concentrated calving of caribou was displaced from industrialized areas to areas of lower forage richness, with caribou returning to industrialized areas during the post-calving period (Griffith et al. ibid). The effects of this herd displacement during calving on population dynamics are debated (NRC 2003). Development conflicts associated with potential habitat loss have been resolved in some areas through "calving group protection measures" (e.g., Canada's Northwest Territories), whereas in other areas (e.g., Alaska and Russia) calving grounds hold no special policy status.

Northern oil and gas development may also influence marine mammals. Noise from offshore oil exploration in the Beaufort Sea disturbs bowhead whales and could deflect them from migration routes, making them less accessible to hunters. Fall migrating bowheads, for example, stay 20 km from seismic vessels (NRC ibid.). Oil spills from marine transportation or offshore oil platforms have the potential for widespread ecological damage, particularly in ice-covered Arctic waters. Spills from pipelines in temperate-zone oil basins in the headwaters of Arctic rivers such as the Ob, Pechora and Mckenzie could also contaminate Arctic waters.

In addition to direct effects and impacts on vegetation and hydrology, oil and gas development has many cumulative effects on the economies and well-being of local peoples, including the impacts of migrant labour, the fragmentation of wildlife habitat, and increased access by non-residents (Chapin et al. ibid., Walker et al.

1987; NRC ibid.). Oil and gas development activities have generally provided few long-term jobs for local residents. However, in North America, where local governments and indigenous organizations that have emerged following the implementation of land claims provide an institutional framework for mitigation and compensation, extractive industries have provided substantial cash infusions to communities in some cases. In Russia, the benefits which Northern Autonomous Districts (such as the Yamal-Nenets Autonomous Okrug) receive from having oil and gas companies on their territories does mean taxes from oil revenues enter the region and constitute a major source of revenue. While many corporations are registered in Moscow, most of the revenues received from strategic resources go directly to the federal centre. Yet Autonomous Districts receive large amounts of investment made by the companies into infrastructure projects. Current legislation requires companies to compensate local indigenous communities on whose lands they operate. In reality, however, this often means individual arrangements are worked out which result in payments and goods delivered directly to the indigenous families working with reindeer or dependent on hunting and fishing. Financial compensation has not been a key means of dealing with local people until very recently. During the early 1990s, anthropologists reported that Khanty reindeer herders and hunters were compensated with snowmobiles, with a few sacks of groceries, with batteries and radios and that, in the YNAO, few compensation payments had been made until recently (Novikova 1995) and some have questioned whether indigenous reindeer herding enterprises are becoming dependencies of oil and gas companies (e.g. Tuisku 2003).

History, Process, Experiences, Perspectives

As interest in the Arctic as one of the world's last energy frontiers increases, this book looks at the emergence of the Arctic as an energy province—in imagination and reality—and, through a discussion of ambitions and plans to explore for oil and gas and to build pipelines—"umbilical cords for the industrialized world"[9]—, it considers a number of case studies from Canada and Alaska, as well as from other circumpolar regions, which illustrate some of the diverse perspectives, interests and concerns of indigenous peoples. The following chapters describe the historical and contemporary interest in northern resource wealth and investigate and examine indigenous and local perspectives on the social impacts of past, current and planned oil and gas activities in Canada's Northwest Territories (NWT) and Yukon Territory, in northern Alberta and northern British Columbia, and in north-east Alaska. They were written against the backdrop of discussion over the proposed Mackenzie Gas Project and the Alaska Highway Gas Pipeline, as well as applications for plans to construct the Northern Gateway Pipeline across northern Alberta and northern British Columbia to transport crude oil produced from northern Alberta's oilsands mines.

Although this book draws on research and travels in north-west North America and other parts of the circumpolar North, as well as the exploration and interrogation of many rich written sources, it does not constitute an ethnography of energy development as it affects a specific community or communities, nor does it report on or attempt to capture the complexity of local dynamics and decision-making processes as they play out in a specific community. It also has its geographical limitations and does not pretend to make a comprehensive circumpolar-wide analysis, nor does it examine and analyze the economics of oil and gas and the economies of indigenous communities. There are many possibilities for this kind of work to be carried out, however. I have written the following chapters as essays or commentaries that consider and reflect upon the idea of the Arctic as a resource frontier and on the concerns expressed by a variety of groups and commentators over the social and environmental impacts of oil and gas development, as well as the opportunities that oil and gas activities will bring to both the long-term viability of indigenous and local communities, and to the sustainability of indigenous and local livelihoods, cultures and societies.[10]

Taken together, the chapters in this book provide a foundation for future work on the political ecology of non-renewable resource development in the circumpolar North, and on the sociology and social anthropology of pipelines (including pipelines as complex and interdependent technological, social, economic and political systems and networks). The broader issues relate to environmental conflict, ecology movements, rights of indigenous peoples, environmental governance, knowledge and power, the social legitimation of megaprojects, sustainable development, energy politics, community empowerment and the role of civil society. While oil and gas exploration is increasing, and the prospects for large-scale development loom large, specific case studies from the circumpolar North remain few and far between in relation to other parts of the world which have seen conflicts between indigenous peoples and oil and gas companies, for instance in South America (see Fontaine 2005, for a review of work carried out in Latin America, specifically the Amazon).[11]

As my interests in the sociology and anthropology of extractive industries are concerned not only with contemporary development activities but with the history of oil and gas activities and mining, part of my focus sharpens on questions that attempt to understand history and process, as well as the experiences that local communities have had with extractive industries, how they oppose or support development, and how various claims for different kinds of knowledge come under scrutiny (e.g., see Gilmartin 2009, for an excellent discussion of conflicts between local residents and multinationals over the construction of a gas pipeline in the west of Ireland). How have parts of the Arctic been regarded by industry, for instance? What influences do the enduring images of the frontier have on attitudes and policy towards circumpolar peoples and lands? And, as industry, government, northern communities and environmentalists await a final decision on the Mackenzie Gas Project, which is expected later in 2010 (possibly coinciding with the publication of this book), what, for example, can communities in Canada's Northwest Territories learn from the experiences of oil development and pipeline construction in Alaska? First Nations in Yukon, anticipating the construction of the Alaska Highway Gas Pipeline, are asking the same question, while First Nations in northern Alberta and northern British Columbia are also keeping an eye on development near their communities and lands as well as being acutely aware that the exploration and exploitation of oil and gas further north will have an eventual impact on them as well. Because the political landscape has changed in many parts of the Arctic, energy companies must increasingly negotiate with indigenous communities and include them in decision-making processes and in environmental and social impact assessments. This may not immediately ease fears of the high environmental and social costs that oil and gas development often leaves in its wake, but it may provide a context for discussion of appropriate strategies for sustainability and environmental protection.

I have made use of a variety of sources, ranging from archival material, published reports, media sources, transcripts from workshops, and the transcripts of public hearings. My work has also been informed by observing and tracking public meetings and regulatory hearings, and my understanding of energy issues has been enriched by discussions and conversations with a number of people, including leaders of indigenous and local communities and energy company officials. Indigenous voices are being heard in the energy development debate. Some are louder than others, while some are muted. Opportunities for the participation of indigenous people in the oil and gas industries have increased significantly in some parts of the circumpolar North in recent years. In Alaska, Canada and Greenland, land claims and the introduction of self-government respectively have provided the means for many indigenous communities to enter into resource development projects through joint ventures with industry and government, impact benefit agreements, and environmental monitoring projects. In northern Canada, comprehensive land claims such as the Inuvialuit Final Agreement (1984), the Gwich'in Comprehensive Agreement (1992) and the Sahtu Dene and Métis Agreement (1994) have given some indigenous people subsurface mineral rights to specific areas of land. In the Northwest Territories, for example, extractive industries such as diamond mining and oil and gas exploration have also provided substantial cash infusions to communities in some cases. Although traditional resource activities and practices remain important to the daily lives of indigenous peoples, the oil, gas and mining industries increasingly provide employment in some areas, especially in Canada and Alaska, and are expected to do so in Greenland and Russia.

Far from being mere victims of the impacts of industrial development, indigenous peoples are participants in, and increasingly beneficiaries of, the development of the Arctic resource frontier. In some cases, they initiate such development. For instance, in Alaska, many Alaska Native corporations have business interests in the oil and gas industry. In Canada's Northwest Territories, a variety of energy-related businesses and companies owned by indigenous organizations and individuals operate out of places such as Inuvik—the Inuvialuit Development Corporation has ownership shares in wells, processing facilities and pipelines, for instance. The Mackenzie Gas Project promises a focus on local Aboriginal involvement, with training initiatives to help the skills and employment needs of northern residents. The Aboriginal Pipeline Group has the right to own one-third of the Mackenzie Valley gas pipeline under a memorandum of understanding (MoU) signed with the Mackenzie Delta Producers Group. In October 2004, Canada's federal government announced Can$9.9 million in funding for the Northwest Territories Oil and Gas Aboriginal Skills Development Strategy, a programme that will aim to provide training for Aboriginal people to find employment in the oil and gas industry. In northern Alberta, Syncrude is one of the largest employers of Aboriginal people in Canada, Mikisew Cree First Nation owns companies employed in the oil industry, and college training programmes in Fort McMurray focus on the trade and heavy industry qualifications that are increasingly required by the oilsands industry. There are possibilities for Aboriginal students to access the funding needed to take such courses through federal and provincial loans and grants, college bursaries or private scholarships. The international resources community has identified the potential for Greenland to be a significant source of new mineral and oil development, with the opening of new mines and heightened interest in oil exploration opportunities offshore of Greenland in recent years. This interest is expected to intensify since the Scottish-based Cairn Energy announced in September 2010 that it had detected the presence of oil in west coast waters, which followed on from an earlier announcement by the company of a discovery of gas. In 2008, the Danish-Greenlandic Self-Rule Commission concluded a series of negotiations on mineral rights, ownership of subsoil resources, and the administration of the revenues from mining and hydrocarbon development. The Commission emphasized that minerals in Greenland's subsoil belong to Greenland and that the country has a right to their extraction. Under the new political arrangement of Self-Rule, which was instituted on 21 June 2009, the Government of Greenland has been granted the rights to administer revenues from the energy and other extractive industries.

Although there is some employment to be gained from existing development, some indigenous communities express disappointment that there is little government support to actually carry out some ventures. In October 2002, the government of the Kwanlin Dun First Nation in Yukon Territory indicated its support for the Alaska Highway Gas Pipeline project, which would pass through some 150 km of Kwanlin Dun traditional territory. It expressed concern that the federal government was only seeing Canada's interest in northern natural gas development as being served by the Mackenzie Gas Project. A desire for oil and gas development comes increasingly from indigenous groups, indigenous organizations and indigenous governments (as we will see in the following chapters, especially in the case of the Aboriginal Pipeline Group in the Northwest Territories). In discussions for the development of a "Pan-Northern Protocol for Oil and Gas Development", First Nations from Yukon, NWT and British Columbia, along with Alaska Native groups, confirmed their support for responsible oil and gas development as long as it respected the land and wildlife and generated meaningful opportunities for indigenous people without compromising their social and economic well-being.

Yet despite the success stories and the ways in which indigenous people can and do participate in the energy industry in parts of the Arctic and sub-Arctic, not all who live on or near the lands and waters where oil and gas exploration and development take place derive benefits from it. Alaskan Inupiat whaling captains, Canadian Inuit hunters and Greenlandic fishers worry about the presence of seismic survey vessels near whale migration routes and feeding grounds and good fishing areas, while hunters and trappers in the Mackenzie Delta and in boreal forest communities are anxious about seismic cuts and pipelines disrupting traplines and traditional hunting lands, as well as the desecration of sacred sites. As will be discussed later on, for example, the Coastal First Nations alliance in British Columbia opposes the Northern Gateway Pipeline project. One of the persistent failings—certainly in Canada—is a process of adequate consultation with indigenous communities and meaningful participation of indigenous people in decision-making processes.

Article 26 of the United Nations Declaration on the Rights of Indigenous Peoples (UNDRIP), which was adopted by the UN General Assembly on 13 September 2007, states that: "Indigenous peoples have the right to the lands, territories and resources which they have traditionally owned, occupied or otherwise used or acquired." Article 32 further asserts that: "States shall consult and cooperate in good faith with the indigenous peoples concerned through their own representative institutions in order to obtain their free and informed consent prior to the approval of any project affecting their lands or territories and other resources, particularly in connection with the development, utilization or exploitation of mineral, water or other resources." This had been earlier emphasized in the World Bank's Extractive Industries Review, yet, around the world, examples of how this does not happen, and how indigenous rights are infringed, are far too numerous. For example, following its acquisition of Burlington Resources in 2006, ConocoPhillips has become one of the most significant international energy companies involved in developing oil and gas resources in the Amazon—in Peru alone, it has drilling rights in areas covering 10.5 million acres of tropical rainforest (Anderson et. al. 2009). In Ecuador, the Shuar, Kichwa and Achuar peoples of the south-eastern part of the country have been calling for the protection of the rainforest in an area now marked off for oil extraction. Elsewhere in the Amazon, indigenous peoples are also attempting to resist the incursions of ConocoPhillips and other multinational companies. In May 2008, the Peruvian government sent troops to back up police who were trying to quell protests by indigenous peoples over land, oil and mineral rights in the Marañon River basin. Indigenous peoples' organizations have accused the Peruvian government of selling those rights to foreign companies and point to a failure to consult with indigenous communities about plans for resource extraction. From the point of view of the government, oil and mineral rights are vested in the state.

Questions are being asked by many communities around the world about the long-term effects of development, on both society and environment, and about what kinds of benefits there will be for local people; in the circumpolar North, indigenous people are asking politicians and industry to state clearly how they will work with them to devise strategies on how the challenges communities face from oil and gas extraction and pipeline construction can be turned into opportunities and how the negative aspects can be mitigated. Indigenous peoples know that the impacts of oil and gas development will be large and, in some cases, irreversible. For example, at public hearings hunters from communities in the northern Northwest Territories express concern that exploratory work has been responsible for scaring away caribou and changing their migratory patterns, with the consequence that local people have to travel further to hunt. At the 11th Inuit Circumpolar Council (ICC) General Assembly held in Nuuk, Greenland in summer 2010, delegates resolved to instruct ICC "as a matter of urgency, to plan and facilitate an Inuit leaders' summit on resource development with the aim of developing a common circumpolar Inuit position on environmental, economic, social and cultural assessment processes and, as a first order of business, raise funds for such a summit."[12] With the news of a discovery of oil in Greenlandic waters, the need for such a summit has assumed an added urgency for ICC Greenland.

The general situation throughout the circumpolar North remains one where indigenous peoples feel they are under increasing pressure to sign up to development projects, to communicate and negotiate with industry and governments, and to adapt to a changing environment resulting from the activities of extractive industries. As a result, some indigenous peoples feel that they are losing control over their homelands and over their livelihoods and are calling for increased participation in consultation and decision-making processes (Nuttall and Wessendorf 2006). They remark how industry arrives in many forms but, whatever the advantages and disadvantages, the protection of future generations should be a priority. These anxieties and concerns were also expressed by Pavel Sulyandziga of the Russian Association of Indigenous Peoples of the North (RAIPON) at the Arctic Leaders' Summit in Hay River in Canada's Northwest Territories in December 2005, and he called for indigenous communities to document their experiences with oil and gas companies. Sulyandziga argued that, despite some stated interests in the protection of the environment and the health of indigenous communities, the reality is that energy companies continue to extract oil and gas and expand their activities in the Arctic, but that many do not acknowledge or respect the rights and interests of indigenous peoples. [13]

The energy industry is also unstable, causes massive environmental damage, and has tremendous social and economic impacts on local communities, as is evident in too many places around the world (think, for example, of the Athabasca oilsands, the Niger Delta, the Gulf of Alaska and the Gulf of Mexico). At public hearings in northern Canada, indigenous people emphasize that they are part of the water and the land, and that they need to protect their future. They point to an urgent need for policy-relevant research into the long-term social and economic impacts of oil and gas development in the Arctic, work that will involve detailed and informed discussion that will contribute to policy recommendations for the amelioration of long-term negative social and economic impacts and consequences.

Acknowledgements

In the course of writing this book, I have received help, support, and constructive criticism from numerous people. At the University of Alberta, my thanks go to Michelle Borowitz, Ryan Brown, Shelby Mitchell and Igor Osipov, who all provided invaluable research assistance and helped me organize a vast and voluminous amount of information and literature on a rapidly changing topic. Funding for travel and research came from the Henry Marshall Tory Chair research programme in the Department of Anthropology and the University of Alberta and the project also draws on research on environmental impact assessment reviews funded by the ArcticNet Network of Centres of Excellence. This book project is also a programmatic activity of the Academy of Finland-funded "Human-Environment Relations in the North" research programme, under the auspices of the Academy's FiDiPro initiative and based at the Thule Institute, University of Oulu, Finland. It contributes to the "Resource Development: histories, patterns, negotiation" component of the programme. At the International Work Group for Indigenous Affairs, I am extremely indebted to Kathrin Wessendorf, who has been tirelessly enthusiastic about the project right from the beginning, as well as being extremely patient as she waited for first drafts of chapters. My work has benefited from her insightful comments and her knowledge of Arctic issues. Also at IWGIA, I would like to thank Cæcilie Mikkelsen for her thoughtful comments, her editorial assistance and excellent work in bringing the book to a final stage of production, and Jorge Monrás for layout and design. I am grateful to Elaine Bolton for her thorough proofreading. My colleagues on the IWGIA Board, both past and present, also deserve a word of thanks for conversation and discussion on a range of issues concerning indigenous peoples and their rights. A special word of gratitude goes to Frank Sejersen at the University of Copenhagen for his careful reading of the manuscript and his suggestions on structure and argument. Finally, my eternal thanks, as always, go to Anita and Rohan.

CHAPTER II

THE ARCTIC ENERGY FRONTIER

The frontier is a compelling image in Canadian historiography and Canadian nation-building. It continues to inform political and cultural ideas and ambitions of economic development and resource extraction at high latitudes, as well as sovereignty and territoriality and the very ideas, images and narratives of Canada as nation, place and space. In North America, the frontier as Frederick Jackson Turner defined it in his address to the American Historical Association in 1893 is usually understood to be an area of free land on the fringes and edges of advancing settlement by pioneers and settlers, the point at which the "wilderness" or "savagery" meets "civilization". In a sense, it is a moving boundary between the settled, the tamed, cultivated and farmed, and the boundless expanses of the wild. It is the frontier, Turner argued, which wielded great influence in the history of the United States and he focused attention on how a geographical periphery could define the character of a nation (Turner 1920). As Walter Prescott Webb put it in his classic work *The Great Frontier*, "The American thinks of the frontier as lying *within*, and not at the edge of a country. It is not a line to stop at, but an *area* inviting entrance" (Webb 1964: 2). Webb argued that the frontier was transient and temporal and "inherent in the American concept of a moving frontier is the idea of a body of free land which can be had for the taking" (ibid.: 3).

Historians have debated the relevance of Turner's frontier thesis for the analysis of Canada's development as a nation. The Canadian Northwest frontier, they point out, differed from that of the American West in that it was not a lawless region populated by hardy pioneers, gunslingers, outlaws and Indians, which was the supposed popular image of the American frontier. For Turner, the American West was a land of opportunity. Yet there was a sense in which Turner was celebrating the open spaces and the wildness the frontier represented and he lamented its loss through the taming of the wild and its settlement and cultivation, whereas Canadian nation-builders saw the potential the vast spaces of the frontier had for immigration and settlement. Its cultivation was something to be celebrated and, in the second half of the 19th century, Canada's Northwest frontier was already being mapped out, demarcated, "tamed", and carefully brought under administrative and legal control before settlers—many of them immigrants enticed from Europe—arrived with dreams of a new life in a promised land.

J.M.H. Careless (1954) advanced a Turnerian "frontierism" argument for how some Canadian nation-builders embraced the idea of the frontier as a way of marking out Canada's distinctiveness from its European ancestral roots. There have been a significant number of critiques, mainly from political ecology and economic geography, about the relevance of Turner's frontier thesis more generally, particularly from scholars and practitioners working in other areas of the world defined as frontier regions, such as the Amazon (e.g. Cleary 1993). Yet despite the arguments about how apt Turner's frontier thesis actually has been for understanding Canada, the frontier has nonetheless stood as a symbolic representation of limitless opportunity, "a metaphor for progress into many spheres transcending physical space" with "the power to shape the imagery of the national character" (Cuba 1987: 155). Like wilderness and borderlines, the frontier with all its spatial, temporal and transitional meanings continues to be fundamental to Canada's geographic imaginary irrespective of its contested nature. For many, frontiers still exist in Canada and their images are perpetuated and entwined with social and cultural identity. As Elizabeth Furniss explores in *The Burden of History* (1999), her rich monograph about the persistence and cultural reproduction of the frontier myth in a rural community in British Columbia, this aspect of the frontier— and as a place at and beyond which what is perceived as seemingly "empty" wilderness remains "untamed" and "untouched"—seems to hold true as much for Canada as it does for the United States.

In 1948, Morris Zaslow concluded that the frontier hypothesis would heavily influence the writing of the future, especially in the ways historians and geographers would look at the shaping of

Canada. Careless (ibid.) reinforced this argument but stressed how the urban, metropolitan eastern Canadian view of the western and northern frontiers was influencing their development. The frontier, he claimed, was shaped by the attitudes and ambitions emanating from the metropolitan core, from where ideas, capital, markets and transportation spread to the hinterland—or what we could also argue could be defined as the geopolitical margins—of the far north-west of Canada. Natural resources played a significant role in this, for it is the case that since the early days of the development of the fur trade, Canada's economy has been founded on and shaped by the production and export of raw materials and the import of manufactured goods. The encounters travellers had with the New World "were shaped by their readiness to find a natural paradise suffused with abundant riches and savage wilderness" (Berland 2009: 86). The frontier remains "in part a metaphor for national development in its material and ideological senses, as well as in terms of spatial expansion and delimitation" (Fold and Hirsch 2009: 95). It is in this vein that I discuss, in this and the following chapters, the images, hopes and ambitions of the development of northern Canada and neighbouring Arctic regions as energy frontiers. In doing so, I draw on both historical and contemporary material to illustrate the processes at play in the transformative effects of resource development on the lives, societies and cultures of northern indigenous peoples and northern environments.

The Arctic is being imagined as a new—although some media and industry commentators are saying last—frontier for oil, gas and mineral extraction, a frontier that is viewed as important for supplying global energy needs and meeting increasing global consumption demands. Oil and gas companies talk of searching for new resources in frontier areas that are harsh and challenging, such as the Arctic and deep water areas. With global climate change affecting the Arctic in an unprecedented way, it is widely assumed that, as sea ice melts and permafrost thaws, access to the Arctic and its resources will be easier in the coming decades than it has previously been in the region's recent history (ACIA 2005). As the world casts its gaze on the circumpolar North for the extraction of resources vital to the functioning of national and global economic systems, as well as its emergence as a region with new shipping routes and opportunities for commerce, scientists, policy-makers, indigenous and local residents and the media all talk about the Arctic in ways that suggest it is on the verge of a transformation into a transnational space firmly embedded in a global economy. Yet this is nothing new to the Far North—regions and people throughout the circumpolar world have a rich history of experiencing the economic, environmental and social impacts of extractive industries.

Mining operations for cryolite in 19th-century Greenland, the Klondike Gold Rush in Yukon at the end of the 1890s, coal mining in the early 20th century in Norway's Svalbard archipelago, oil production at Norman Wells in 1920's Canada, High Arctic oil and gas projects in the 1960s—to say nothing of the major extractive industries developed in northern Fennoscandia and Russia throughout the 20th century—are examples of the thousands of capital-intensive and scale-expansive operations that have happened all over the world, and which continue to expand, operations that Bunker and Ciccantell (2005) say have moved the global economy towards greater globalization. They argue that globalization is the latest manifestation of capital's internal dynamic and that it results from processes of material and spatial expansions and intensifications which are driven by economies of scale made possible by technology. What we are witnessing in the Arctic and other regions that are defined as the world's last frontiers, and which are sought out by the transnational players constituting the world's extractive industries, is merely the latest chapter in a "historically constant process of expansion". Bunker and Ciccantell (ibid.: xiii) suggest that capitalism is deeply rooted in "the ongoing, cumulatively sequential expansion of its own reproduction". However, it is a process, they argue, which may be reaching its global limits—quite literally in a geographical, ecological and material sense—as experienced in the Arctic, the Amazon and other remote regions of the globe. Frontiers are broken down, geographical space runs out and resources are used up. It forces us to question the development discourse of limitless horizons and boundless opportunity.

De Angelis (2004) has argued that resource frontiers are essential spatial forms for the successful functioning of global capitalism, while Walker (2006) sees frontiers as expanding borderlands driven by economic investment and development, which may turn out to be short-term and are characterized by cycles of boom and bust. For many indigenous peoples around the world, from the Arctic to the Amazon, from remote mountain valleys to tropical forests, deserts and tundra, globalization is regarded as a process of cultural homogenization which entangles local cultures in a struggle with global forces. This perception of globalization may emanate from their experience of extractive industries, oil and gas companies and mining ventures, which are some of the most visible and tangible aspects of this worldwide process.

Indigenous peoples are often engaged in struggles with regional and national governments—as well as with extractive industries operating on or near their lands—which are ultimately about being able to maintain community survival, cultural diversity and indigenous livelihoods, gaining recognition of cultural and political rights, and asserting ways that ensure cultural protection in the face of threats to cultural and economic survival. Deriving benefits from, as well as participating in the extractive industries and gaining a measure of positive economic development for their families, households and communities are also stated objectives for indigenous peoples in many cases. While access to new goods and other cultural items, and the availability of services such as education and healthcare, are often welcomed by indigenous peoples, and while globalization and resource development bring new opportunities and open up exciting vistas, nonetheless there are concerns that globalization, through the exposure it gives to foreign societies, goods and cultural values, threatens the viability of local languages and dialects, of traditional value systems and ways of life, customary resource and land use, and locally-made products and the people who produce them. This is particularly the case in areas defined as resource frontiers and which are affected by the influx of migrant workers—such as those needed for mining operations and the construction of pipelines—and boom and bust cycles of development which have marginalized indigenous people and challenged or even eradicated customary regimes of property and land use (Nuttall 2010).

The destruction of indigenous livelihoods and local environments is often connected to industrialization and modernization, a process noted by the Brundtland Report, which emphasized that traditional lifestyles of people around the world are "threatened by insensitive development over which they have no control" (WCED 1987: 12). Many critics of globalization also argue that its effects can be felt in the corporate exploitation and theft of traditional knowledge and intellectual property rights, often framed as an infringement of human rights, cultures and ecosystem biodiversity. Bunker and Ciccantell's argument is that globalization is a corporate and economic elitist view, a perspective on the world that sees resources as there for the taking by the most competitive and most powerful (2005).

19th Century Canadian Frontiers

The Dominion Lands Act of 1872 set in motion the topographic reshaping and the demarcating of administrative boundaries of the Canadian West by surveying and dividing the land into quarter sections ready for the backbreaking work of homesteading. In 1873, the North-West Mounted Police was established to pave the way for peaceful settlement and, in the same year, the Department of the Interior was set up to devise, amongst many other things, methods of removing and clearing indigenous Indians from the plains and prairies and settling outstanding grievances with Métis groups as well as actually administering the disposal of acres for homesteading and encouraging immigration. Thus land surveyors, treaties with Indians, the granting of scrip to Métis, the implementation of law and order and the construction of railways all preceded the pioneers who came to cultivate the landscape and fashion new forms of social life in pockets of settlement surrounded by vast wilderness. Berland (2009: 21) argues that Canada has its own history of dispossession in the making of the frontier, with which the narrative logic of Canadian nation-building is complicit.

Analyzing public museum displays about the lives of the pioneers and settlers in Williams Lake, British Columbia, Furniss (1999) notes how Aboriginal people provide a historical background to settlement—they are imagined and depicted as living quietly and passively in the forests and fishing the rivers and streams of the wilderness, but not considered active participants and historical agents in turning the land from empty wilderness to cultivated civilization and, in doing so, helping to construct the nation.

As the frontier was pushed back further west in the 19th and early 20th centuries, as a railway was built connecting eastern Canada to the Pacific coast of British Columbia, and as immigrants from Europe settled the prairies, and mining towns sprang up on the Canadian Shield and in the foothills and valleys of the mountainous west, the last remaining areas for settlement and development were now supposed to be found only in more northerly parts of the country. The western frontier thus moved north but, as Morris Zaslow (1971: xi) argued, "'North-West' and 'North' are more than geographical expressions, they constitute a process: the advance of frontiers and frontier experience from the rear of the Province of Canada to the prairie northwest, then gradually northward along several fronts to the northern coasts of Canada and the islands beyond. Such frontiers were of many kinds—of societies, cultures and administrations, as well as of industries and people."

By the last two decades of the 19th century, the potential for settlement and agricultural development in the upper Peace River region of what was to become northern Alberta, and even north of 60° latitude, was generating tremendous excitement (up to this time, Edmonton was considered the extreme limit of settlement). Following the signing of Treaty 8 in 1899 (see Chapter Three), it was beginning to look like a reality. Zaslow (1971: 201) cites reports in the early years of the 20th century from the Dominion Lands agent in Edmonton, who claimed that the region west and north of Edmonton—the Peace River area and the Mackenzie River Basin—could hold two million people. Journeying from Winnipeg through north-west Canada via Edmonton to the Mackenzie Delta and Arctic coast in 1908, Agnes Deans Cameron observed that *the West that we are entering upon is the Last West, the last un-occupied frontier under a white man's sky. When this is staked out, pioneering shall be no more, or Amundsen must find for us a dream-continent in Beaufort Sea* (1986 [1909]: 3).

Heading north of Edmonton with her niece as travelling companion, Cameron reflected on the inevitability of the expansion of settlement into the "new North" and wanted to meet the "Trail-Blazers of Commerce, who, a last vedette, are holding the silent places, awaiting that multitude whose coming footsteps it takes no prophet to hear" (ibid.: 2).

Late 19th and early 20th century pioneering was aimed at breaking land and pushing back the northernmost frontier of wheat growth. Experimental agricultural stations were established in the Mackenzie Basin and the Yukon under the auspices of the Experimental Farms System of the Dominion Department of Agriculture. Conducting exploratory surveys of soil conditions in northern British Columbia, Alberta, Yukon, and in areas along the Mackenzie River in the 1930s and 1940s, officials from the Experimental Farms System were confident about the possibilities of agricultural development in the north-west and even as far north along the route between Whitehorse and Dawson City in the Yukon. Yet they were also realistic in their assessment of the Mackenzie Valley as a place that would probably not, in the foreseeable future, offer a field for agricultural development. Reports concluded that distances were too great, the environment far too rugged, and potential markets too small to justify agricultural production as anything more than being subsidiary to other viable enterprises, namely the fur trade, oil and mining.

Despite the climate, agriculture in Canada's northern regions may not be such a fanciful idea—crab apples ripen beside Great Slave Lake, vegetables have been cultivated on the Arctic coast, and potatoes have been grown close to the Arctic Circle. According to the Arctic Council's Arctic Climate Impact Assessment, northern regions will likely see the tree line expand north and, with changes in vegetation and soils under conditions of climate warming, the potential for commercial crop production is projected to advance northward throughout this century (ACIA ibid.). Yet, irrespective of historical documents or scientific scenarios raising discussion of the prospects for cultivation of land north of 55° in Canada (farming in northern Alberta, which is carried out up to a latitude of 58° 24' at Fort Vermillion, is considerably far north by Canadian standards), it has been the North as a mineral and hydrocarbon-rich resource hinterland that has long been a mainstay of Canadian mythology about the frontier and ambitions for its development (Dacks 1981). While the Gold Rush was transforming the Yukon over 100 years ago, a Canadian Geological Survey publication of 1898 reported gold at Yellowknife 40 years before the first ore was mined, oil seepages in the Mackenzie Valley were recorded more than a century before the first wells were drilled at Norman Wells on the banks of the Mackenzie River, and the first strike of paying qualities of oil was made in the Fort McMurray area of northern Alberta in 1909, leading to the leasing of lucrative parcels of oil-sands land.

The Trail North

With the exception of the vast Athabasca oilsands mining operations in northern Alberta, there has been a lack of large-scale extraction of hydrocarbons in Canada's North relative to the kinds of oil and gas development operations seen in some other parts of the Arctic and sub-Arctic, such as in Alaska and Russia. But this is not to say that oil and gas exploration and speculation does not have a long history in northern Canada, or that there have not been booms in oil and gas production. Explorer and fur trader Peter Pond was the first European to see the oil-saturated deposits of sand now so important to Alberta's economy, although reports date from around 1719—some 60 years before Pond not only saw it *in situ* but also witnessed indigenous people using the tar to waterproof their canoes—that Cree traders had taken bitumen to the Hudson's Bay Company post at Fort Churchill on the western shores of Hudson Bay. The traders at the fort did not know what to do with it—its worth or its wider use could not be imagined—and they were more interested in obtaining the furs the native people had brought from the north-west interior. In his journals, another European explorer, Alexander Mackenzie, described the Athabasca oilsands, evidence of which he had seen a few years after Pond in 1788. He came across

> some bitumenous fountains, into which a pole of twenty feet long may be inserted without the least resistance. The bitumen is in a fluid state, and when mixed with gum, or the resinous substance collected from the spruce fir, serves to gum the canoes. In its heated state it emits a smell like that of sea-coal. The banks of the river, which are there very elevated, discover veins of the same bitumenous quality.[14]

Further north, in 1789, in the valley of the river that was to bear his name, Mackenzie was the first European to notice oil seeping from the ground around the area that is now known as Norman Wells. Yet, before Mackenzie's journey and the arrival of fur traders to the region, indigenous Dene who lived along the Mackenzie River—or the *Deh cho* ("great river") as they called it—did not need European explorers to tell them about oil in the area. Like the Cree further south, they gathered tar from the bituminous seeps and mixed it with tree sap to make a waterproofing substance for their canoes. It is likely they also considered it to be a resource of cultural and economic importance and used it as an item of trade in their dealings with other indigenous groups before European contact. Fur traders later recorded in their journals that tar from Rond Lake near Fort Good Hope was scooped into buckets and traded with other Hudson's Bay Company outposts. It was Dene guides who, in the early 1900s, worked with geologists who were exploring the region around Fort Norman (Tulita) and took them to Legohli, a place meaning "where the oil is" in the Dene language.

In the late 1880s, and having confirmed Alexander Mackenzie's reports from the end of the previous century, the Geological Survey of Canada was confident about the great potential of the Mackenzie oil seeps, although the remoteness of the region was seen as an impediment to any immediate development. In 1914, T.O. Bosworth, a British geologist, was commissioned by two Calgary businessmen, F.C. Lowes and J.K. Cornwall, to assess the petroleum potential of northern Alberta and further north in the Mackenzie Basin (McKenzie-Brown 1998). He set out from Edmonton in May 1914, returning four months later. During his expedition, he investigated the oil seeps and geological structures that were already known to hold possible reserves and concluded that the prospects for exploration were excellent in three main regions: the Mackenzie Valley between Fort Good Hope and Fort Norman, Great Slave Lake where tar springs were found, and the oilsands district of the Athabasca River.

In his book the *Richness of Discovery*, historian Peter McKenzie-Brown sees Bosworth's expedition as pivotal for the beginning of western Canada's oil industry. Bosworth's "Report on the Prospects of Obtaining Oil in the Regions of the Mackenzie River, Great Slave Lake, Slave River and Athabasca River", written shortly after his return from the North, was influential in generating the interest which led to the eventual economic development of the Norman Wells area. This began when Imperial Oil Ltd began exploratory drilling for oil in 1919 (having acquired the claims staked by Bosworth) and opened a refinery in 1920 following the discovery of oil that same year. Imperial determined that the same kind of Devonian geological structures they were drilling into at Norman Wells existed further south and this led to the development of the major oil field at Leduc in Alberta in 1947.

Despite the Norman Wells venture, exploration and drilling continued at low levels in the NWT, but threats of Japanese attacks on Alaska during World War II spurred the American government to initiate the Canadian Oil (Canol) Project. To supply the military build-up and infrastructure necessary to protect the coast, the U.S. required ready access to oil. The Canol Project involved constructing a pipeline from the oil fields in Norman Wells for 925 km across the rough terrain of the Mackenzie Mountains, a divide separating the watersheds of the Mackenzie and Yukon rivers, to a newly constructed refinery in Whitehorse in the Yukon. From there, oil was transported by a network of pipelines to points along the Alaska Highway, including to a fuelling station in Skagway in south-east

Alaska. The pipeline was constructed rapidly during 1943-44 by civilian contractors employing both Canadian and American workers. The urgency of getting the job done was reflected in the poor quality of the end result. Sections of pipeline were laid on the surface of the ground and crude oil frequently leaked into the permafrost. During its first nine months of operation, for instance, some seven million litres of crude oil were spilled along the length of the pipeline. Although seen as a major event in the history of Canadian cold region engineering, it was nonetheless designed, constructed and operated with very little understanding of the need for unique northern design and construction methods.

For 16 months, from mid-December 1943 to the beginning of April 1945, crude oil was pumped through the pipeline, although it took from December 1943 to April 1944 for the first oil to actually reach Whitehorse. At the peak of production, 4,400 barrels of oil a day were passing through the pipeline to Yukon. But it was an expensive, short-lived project, what some called a colossal blunder. In March 1945, 11 months after the oil first arrived in Whitehorse, the U.S. Army issued an order to stop its flow through the pipeline. The project was terminated and, after the end of the war, some sections of the pipeline were dismantled, and the refinery at Whitehorse was closed down (it had processed 156 millions litres of crude oil from Norman Wells during the lifetime of its operations). At the time, it was one of the largest projects ever undertaken in northern Canada and the major infrastructure was not only the Canol pipeline itself but over 1,500 km of other subsidiary pipelines (including a distribution line from Whitehorse to Fairbanks), almost 1,000 km of gravel roads and around 2,400 km of winter roads. In addition, a telephone communication system was constructed between Norman Wells and Whitehorse.

The environmental legacy of the Canol Project remains very much in evidence. Although Imperial Oil acquired the rights to salvage the decommissioned and defunct equipment and facilities of the project in 1947 (the same year it had discovered the Leduc oil field south of Edmonton), no real attempt was made to completely deal with any of its environmental impacts. Salvage crews moved in along the route and took away power units, motors, pieces of pipe and brass valves. The Whitehorse refinery was not broken up for scrap. It had cost Can$24 million to ship the refinery from Texas at the beginning of the Canol Project and Imperial, which had purchased it as part of the salvage operation, decided to move it to Edmonton. Today the pipeline route is still lined with abandoned camps and shelters, pumping stations, vehicles and other equipment and, in 1998, some sites of the Canol Project were declared to be environmentally contaminated. Much has been cleaned up since and parts of the route have been designated a heritage hiking trail, which attracts backpackers in search of a wilderness adventure.

While exploration and drilling for oil and gas took place both before and after World War II, the main cycles of activity occurred from the 1960s to the 1980s. For example, the first well to be drilled in the Canadian High Arctic was on Melville Island during the winter of 1961-62 and a significant oil discovery was made at Bent Horn on Cameron Island in the early 1970s. This find confirmed the potential for even more oil and gas to be present in Canada's far northern and offshore basins, which have thicker and younger sedimentary rocks. Operated by Panarctic Oils, the first shipment of 100,000 barrels of Bent Horn oil was made in 1985 by the M.V. *Arctic*, a specially reinforced ice-breaker tanker to a refinery in Montreal and shipments continued until the late 1990s. The development was not without controversy. As Jull (1990) observed, Inuit were concerned at the way Panarctic tried to deal directly with Inuit communities instead of going through territorial and federal administration and regulations, and wanted a proper consultation process, which many felt was absent.

Oil and gas production in the Arctic prior to 1970, however, was small-scale and the oil and gas recovered was mostly shipped to southern markets for refining and distribution. This shipping was restricted by the extreme climate of the Arctic—sea ice affected maritime routes but transporting large quantities of oil south was also an extremely expensive business. Exploration offshore in the Beaufort Sea began in shallow water leases in the mid-1970s. In 1976, the Canadian government approved Dome Petroleum's application for licenses to drill in the Beaufort Sea and the federal budget of the following year introduced the Frontier Exploration Allowance, which provided tax allowances to Beaufort Sea operators. In 1978, the Endicott oil field (with estimated recoverable reserves of 500 million barrels) was discovered near the Sagavanirktok River Delta east of Prudhoe Bay in Alaska, and this still makes energy producers confident that similar finds will be made in the Canadian section of the Beaufort Sea. BP and Exxon recently spent Can$1.2 billion and $585 million respectively on exploration licenses for use in Canadian Beaufort waters. The area, however, is a contested one. There remains disagreement between Canada and the U.S. over how the maritime boundary in the Beaufort Sea between the two countries should be agreed upon and drawn. Claims for the ownership of over 6,000 square km of ocean, which could reveal some of the best prospects for oil and gas discoveries, remain unresolved.

The first major northern energy megaproject in the NWT to receive investment was the Norman Wells Project, carried out between 1982 and 1985. Apart from the increased development seen during the Canol Project, whatever oil had been produced from Norman Wells had, until then, been supplied to local communities and mines along the Mackenzie River and on the shores of Great Slave and Great Bear lakes but, with oil prices rising, Esso Resources Canada saw an opportunity to expand production (Bone (ibid.: 2009: 178). A pipeline was built to Zama, Alberta to allow for easier transport of oil to southern markets. Norman Wells would subsequently become the third most productive field in Canada, and became Imperial Oil's largest source of crude oil. Although there are wells in the actual Norman Wells townsite, much of the oil is produced from wells which have been located on artificial islands in the Mackenzie River or along the banks of the river. Bone (ibid.: 179) argues that industry and government saw the Norman Wells Project as a successful model for how future energy development should proceed in Canada's North. Oil executives applauded the project because it had been completed without any serious environmental impact; as a feat of engineering and technology, it was also the first pipeline in Canada to be completely buried in the zone of discontinuous permafrost.

When approving the project, the federal government had stipulated that emphasis be placed on involving northern companies and hiring northern workers and it was argued that an indicator of the success of the project was the fact that the community and residents of Norman Wells benefited economically. The pipeline had some positive economic effect on the communities along the route, but most of the financial impact was felt in Norman Wells. With the rapid development that accompanied the Norman Wells Project, the federal government also provided funding for social programmes, such as housing and dealing with social problems, which were intended to minimize any negative social impacts on nearby communities. Yet as Gorman (1997) shows, community members responsible for organizing the programmes encountered administrative problems in the application process which resulted in delays in the delivery of funds for their implementation. Dene community members also felt they did not have the opportunity for meaningful input in the decision-making process for how the social programmes would be determined and operate. Once initiated, the main focus of the social programmes was on drug and alcohol abuse, and they were discontinued once funding was exhausted.

At the time, the Norman Wells project was considered an unexpected success story for future pipeline development because it was initiated after the Berger Inquiry (see further below) and before the negotiation and settlement of Aboriginal land claims. Interest in the potential of vast Mackenzie Delta and Beaufort Sea reserves had heightened in the 1970s. Substantial discoveries of oil and gas had been made and the Geological Survey of Canada estimated that between 9 and 12 billion barrels of oil and as much as 4.1 trillion cubic metres of natural gas were to be found in the region. It was questionable how much of this was all that accessible, but industry considered that a good part of the estimated oil and gas reserves was probably marketable and a proposal to construct an oil or gas pipeline along the Mackenzie Valley to northern Alberta became a significant issue of public policy. In 1974 the Canadian Arctic Gas consortium submitted a formal application to the Canadian government to build a natural gas pipeline, and a Royal Commission of Inquiry was established to assess the potential environmental, social and economic impacts. It was chaired by

Justice Thomas R. Berger and the legacy of his inquiry is discussed in Chapter Three.

The Modern Resource Frontier

The exploration for, and ultimate exploitation of, oil and gas in the Arctic began in the 1920s but much of the rapid expansion occurred during the second half of the 20th century. Writing in 1973 at the height of the energy crisis, Richard Rohmer reported how

> *A slumbering, frozen giant is coming alive. Canada's last frontier – the Arctic – is emerging with enormous strength, power, and rapidity. In the short space of four years, it has become one of the major natural resource areas of the world, and is now capable of either gripping Canada by its economic throat or, if controlled, of giving Canada a guiding hand into a prosperous future.* [15]
> (Rohmer 1973: 8)

In Canada, plans for the development of minerals and oil and gas have moved higher up federal government agendas in recent years, as the country appears to be making a transformation into a petro-state (or an "emerging energy superpower" in the words of Prime Minister Stephen Harper) and as international concerns over climate change, environmental protection, mineral exploitation and energy security converge to make the Arctic a geopolitical hotspot. By the first decade of this century, natural gas accounted for around 40% of Canadian primary energy production, followed by oil at 24% (Pratt 2001) and both government and industry consider the development of Arctic resources to be critical for future supplies. Oil and gas activities are major drivers of social and economic change in many parts of the Arctic, not just in the Canadian North. Exploration and development look set to expand and the potential for the Arctic to become a major hydrocarbon-producing region is becoming a significant economic, geopolitical and societal issue (Nuttall and Wessendorf 2006, Mikkelsen and Langhelle 2008).

At the close of the first decade of the 21st century, oil and gas companies are talking about the Arctic as a last frontier for energy exploration and development. Officially in Canada, "frontier lands" are defined under the *Canada Petroleum Resources Act* as lands that belong to Queen Elizabeth II, as head of state in right of Canada, or in respect of which the Queen in right of Canada has the right to dispose of, or to exploit the natural resources of, those lands, although strictly speaking this applies to Nunavut and the Northwest Territories but not to Yukon because of recent devolution agreements. Frontier, however, has become a catch-all word to capture the image of Arctic lands and seas as places at the edge of current development, places which are awaiting exploration but which are at the farthest remove from the core, places inviting entrance, to draw on Walter Prescott Webb's idea. As a distant place, the frontier is remote but nonetheless retains a unified relationship to the core and stands as the symbolic representation of limitless opportunity, at once a place where a bounty of economic resources is to be found, and a repository of cultural symbols (Cuba 1987: 154-5).

At the Inuvik Petroleum Show in summer 2009 (an annual event which takes place in the Mackenzie Delta town), a senior official with Imperial Oil spoke of the Beaufort Sea and other deep Arctic waters as "new frontier ground". Lining up to talk about their plans for the next four to five years, executives of ExxonMobil and BP, as well as Imperial, seemed to suggest that a deep water drilling boom was about to take place in the Arctic. This is one part of the world where it has not been easy to look for hydrocarbon deposits, yet energy companies now see deep water oil, not just in the Arctic but in other areas of the globe, as the real frontier of the future and the deep water areas off Arctic Canada, Greenland, Norway and Siberia, as well as the Gulf of Mexico and the coasts of Africa and Brazil are the places where most oil companies expect to find the bulk of the world's undiscovered oil. In these areas, seismic surveys have revealed subterranean structures which closely resemble those beneath the oil-rich North Sea.

While the Canadian Arctic has not yet experienced the same scale of energy and mining megaproject development as found in some parts of Canada's sub-Arctic (including the accompanying and often profoundly disturbing social and environmental impacts), much of the country's undeveloped oil and gas potential lies in the Far North. Spectacular gas discoveries in the Northwest Territories and the prospects of further finds elsewhere in the North have reinforced the idea of this huge part of Canada as a future "resource hinterland interlocked into the global economy" (Bone 2003: 103), a region critical for the development of the country as a whole but with specific challenges and opportunities for Northerners. The North, for better or worse, is teetering on the verge of a major resource boom and the supporters of oil and gas development and mining projects not only argue their case in terms of economic development but within the context of Canadian national ambition and sovereignty.

This was underscored in August 2008 when, ahead of a three-day visit to northern Canada, Prime Minister Stephen Harper announced funding of Can$100 million to initiate a detailed geo-mapping programme for minerals and oil and gas deposits. As part of the government's Northern Strategy announced in October 2007, the Geo-mapping for Energy and Minerals (GEM) initiative is to be implemented by the Geological Survey of Canada and is a scientific and technical programme designed to deliver geoscience knowledge specifically for economic reasons, and to attract private investment. The Northern Strategy envisions a new kind of Canadian North (Kozij 2009) and makes the region one of the Canadian government's top priorities. The October 2007 Speech from the Throne, which is the federal government's expression of purpose, announced that the government would bring forward an integrated Northern Strategy based on four themes:

- Sovereignty - protecting Canada's Arctic sovereignty as international interest in the region increases;
- Economic and Social Development - encouraging social and economic development and regulatory improvements that benefit Northerners;
- Environmental Protection - adapting to climate change challenges and ensuring sensitive Arctic ecosystems are protected for future generations; and,

- Governance - providing Northerners with more control over their economic and political destiny.

The GEM initiative will map and chart areas of the North with high resource potential that will help energy and mining companies target new exploration sites. Phrased within the context of a new assertion of Canadian sovereignty in the Arctic and that climate change presents not just threats but opportunities, Harper's announcement expressed the Canadian government's view that the development of energy and mineral resources is the primary source of economic growth in Canada's North, leading to new job opportunities and contributing to community viability, as well as helping Canada to emerge as an "energy superpower" (a point Harper has made a feature of his foreign speeches). Mary Simon, President of Canada's national Inuit organization, Inuit Tapiriit Kanatami (ITK), was reported in the press as welcoming the GEM initiative. When Harper reached Inuvik on 28 August, he used the occasion of his Arctic tour (as well as the 50th anniversary of Inuvik's creation as a planned modern town on the Mackenzie Delta) to announce that his Conservative government had commissioned a new polar-class icebreaker, to be christened the *John G. Diefenbaker*, after the leader of the Progressive Conservative Party who was Prime Minister from 1957 to 1963.

Diefenbaker had a vision for Canada's North and he made the region a cornerstone of his re-election campaign in 1958. Perhaps positioning himself to be remembered as a new Diefenbaker, someone who did not ignore Canada's northern reaches and someone whom people would look back on as having left a legacy of positive development and economic improvement for northern communities, Harper has focused on Canadian sovereignty over the Northwest Passage and the High Arctic Archipelago, including strengthening Canada's military and scientific presence in the North, and has expressed a commitment to building sustainable communities and developing northern resources. Diefenbaker made roads a central, but unexpected, part of his vision for Canada and its North. Diefenbaker looked back to the achievements of John A. Macdonald, Canada's first Prime Minister, referring especially to his efforts in opening up Canada's West. "We intend to start a vast roads program for the Yukon and Northwest Territories," he said in the opening speech of his election campaign,

> *which will open up for exploration vast new oil and mineral acres – thirty million acres! We will launch a seventy-five-million dollar federal-provincial program to build access roads. THIS IS THE VISION…We are fulfilling the visions and dreams of Canada's first prime minister – Sir John A. Macdonald. But Macdonald saw Canada from East to West. I see a new Canada. A CANADA OF THE NORTH!*
> (Quoted in Coates and Morrison 2005: 284)

Once in office, Diefenbaker committed Can$100 million for investment in a northern infrastructure programme which became known as "Roads to Resources"—over 2,000 kilometres of roads were built across the northern territories, yet ambitions for other roads, including one along the Mackenzie Valley, were eventually unrealized. The "Roads to Resources" programme was a strategy for developing Canada's North and linking it to the south but the importance of the North must be seen with reference to Diefenbaker's push towards economic nationalism. As Coates and Morrison (1992: 87) have pointed out, the "press and the public responded with enthusiasm to his call for people to build on the country's essential nordicity, to capture the northern Eldorado, the riches which lay between them and the Pole."

In Inuvik, 50 years later, Harper placed the North at the very centre of government strategy for economic development, arguing that the true North was Canada's destiny and that to ignore its promise and ascendancy would be tantamount to Canadians turning their backs on what it means to be Canadian. In a sense, Harper has launched a "maps to resources" programme, setting out a strategy for how and where to locate minerals, energy and gas, and a charter for infrastructure development to get those resources out of the ground. Roads, waterways and harbours are important elements of this strategy. Roads, as in Diefenbaker's government, are important aspects of political economy and ideology. Roads go in the direction of development; they are crucially important arteries that connect people and communities and allow for the flow and movement of people between places, as well as allowing access to regions and resources previously considered inaccessible. Roads are also important statements of political ideology, used to inscribe territorial claims in the landscape, break down barriers of remoteness and push back the frontier.

In some ways, the Northern Strategy is a response to decades of government neglect of Canada's northern territories and its peoples. But it is also a political response to the effects of Arctic-wide social and economic transformation, to environmental issues resulting from climate change, to growing international interest in Arctic resource exploration and development, and a reaction to continuing international opinion (expressed mostly by the United States) that the Northwest Passage is an international strait, not internal Canadian waters subject to Canadian jurisdiction. Also in August 2008, Canada announced it was to tighten its shipping regulations in the Northwest Passage, with mandatory registration of foreign vessels sailing in Canada's Arctic waters. The U.S. response was swift, and stated that the Canadian move would be reviewed to ensure it was consistent with the international law of the sea. In November 2009, Canada's House of Commons began a debate on renaming the Northwest Passage the "Canadian Northwest Passage" as an attempt to affirm, albeit symbolically perhaps, the country's ownership of waters which Canada is well aware may become increasingly disputed in the near future.

Sovereignty disputes, the determination of the extent of Arctic borders and exclusive economic zones (EEZs), and the legal entitlement to use certain Arctic maritime passages and seaways have come to dominate some discussions over Arctic governance and Arctic resources. A Russian government security report released in May 2009 predicted that military conflict over Arctic resources was possible. This prompted Canada to release a statement saying that it would step up its presence in its own Arctic and would not hesitate to defend its interests in the North (Dey Nuttall and Nuttall 2009). In a speech delivered at the Economic Club of Canada in Toronto at the end of November 2009, Foreign Affairs Minister Lawrence Cannon said he was sending a "clear message to the world" that, while Canada would work in a cooperative way with other Arctic nations over the future of the circumpolar North, it would always stand up for its interests and ownership over the Arctic. Cannon made explicit reference to the idea of the Arctic as an energy frontier, saying that Canada's future as an "energy superpower" was dependent on the potential exploitation of the rich deposits of oil and gas on land and seabed (Boswell 2009). While Cannon's comments were directed at other Arctic nations, they were aimed specifically at Russia. Yet the delivery of his speech coincided with news that Canada had lodged a protest with the U.S. government over plans to auction oil and gas rights in an area called Tract 0001, which falls entirely within a disputed section of the Beaufort Sea that is contested by both countries, as well as with the launch, by the U.S. Navy's climate change task force, of a document to help guide its activities and operations in a future ice-free Arctic.

In July 2009, on the occasion of the appearance of a report called *Canada's Northern Strategy: Our North, Our Heritage, Our Future*, the Canadian government announced it was launching a major public relations campaign in support of its assertions of Arctic sovereignty and its claim to be a leader in international circumpolar affairs. As John Kozij points out,

> *Sovereignty and security has been the pre-eminent signature of this Government's vision for the new North. At a glance one can see investments that indicate Canada's strengthened presence on the land, sea and sky over the Arctic. These investments are important because increased traffic in the North can have both positive and less positive elements and an increased presence in the North and the icy waterways of the Arctic is in all nations' best interest* (2009: 13).

Arctic politics has been placed firmly within the Canadian government's election strategy as well as its domestic and foreign relations policies. Critics of Canada's awakened concern with the North point out that the interests of the indigenous peoples of the region, however, only seem to be of concern when Canadian sovereignty appears to be threatened or when resource development promises sound returns on the investment put into it. As Abel and Coates (2001: 12) have written, "the lack of vision or imagination was a mixed blessing for northern Aboriginal peoples. Certainly, it kept the Canadian state from interfering in their lives for much longer than might otherwise have happened, but at the same time, northerners were unable to get the attention of the state at times of crisis when help was genuinely required." Inuit leaders have argued that sovereignty issues offer an opportunity to include indigenous perspectives on Canada's North. For example, they insist that an Inuktitut name for the Northwest Passage—*Tallurutik*—would be appropriate.

Looking North: the Last Frontier?

The Mackenzie Delta and Mackenzie Valley in the Northwest Territories, in particular, have re-emerged and been re-imagined in recent years as energy frontier regions critical for the economic future of Canada's northern territories, but also for the nation itself. But the transportation of Canadian energy supplies does not stop at Canada's southern border. Corporate and political leaders in the United States and Canada imagine a future of energy interdependence on the North American continent, with northern Alberta's oilsands as a hub for further development and an anchor for integrating all North American energy supplies. Exports of Canadian gas "have been the most dynamic factor" in the expansion and growth of "an integrated continental market in energy" (Pratt ibid.: 10) and, as Nikiforuk (2008: 183) writes, "Continental integration assumes that longer global supply lines for hydrocarbons are sustainable and that Canada has cheap energy to spare." In this vision of the energy future, oil and gas pipelines extend further and further throughout Canada and the United States in an attempt to ensure and secure continental energy autonomy through energy interdependence.

Based on available seismic data and current understanding of subsurface geology, it has been assumed for some time that much

of Canada's undeveloped oil and gas potential lies in its northern Arctic and sub-Arctic areas. The NWT and Nunavut host an
estimated 33% of Canada's remaining conventionally recoverable
resources of natural gas and 25% of the country's remaining recoverable light crude oil. Running between the Rocky Mountains and
the Canadian Shield, about half of these potential resources lie in
the western Arctic, making them strategically located—and therefore more accessible—north of existing production infrastructure
in the western provinces of Alberta and British Columbia (INAC
2005). Geologically, the lands extending north of 60° from the
Alberta-British Columbia border through the Mackenzie Valley to
the Beaufort Sea are a continuation of the oil and gas-rich Western
Canada Sedimentary Basin. Added to this are potential reserves
in Yukon Territory, although exploration and development there
have not reached the same levels as seen elsewhere in the North.

Over the last decade, energy companies have begun to look further northwards in their search for viable hydrocarbon reserves. In
the early to mid-2000s, renewed interest in prospects for oil and
gas exploration in northern Canada followed on from increasing
continental demand, pressures on supply and the depletion of
major gas reserves throughout North America, and a rapid rise in
energy prices. Innovations in the geotechnical, ice engineering and
pipeline construction fields have also been instrumental in opening up Arctic frontier regions for exploration and development.
But, as will be discussed later, favourable signals from Aboriginal
communities, corporations and political leaders towards energy
development and pipeline construction have also guided energy
companies towards thinking about the North as a place to focus
investment. Several large-scale development applications have recently been submitted to Canadian federal, provincial and territorial regulators. The most ambitious, but perhaps also the most controversial, development plan pending final regulatory approval is
the Mackenzie Gas Project (MGP), a Can$16.2 billion joint proposal
by Shell Canada Limited, ConocoPhillips Canada (North) Limited, ExxonMobil, Imperial Oil Resources Ventures Limited and the
Aboriginal Pipeline Group (collectively referred to as "the proponents"). This is a megaproject that would see the development on

(mainly) Aboriginal lands of natural gas from three fields in the Mackenzie Delta area for delivery to markets in Canada and the United States, as well as to power further development in Alberta's already massive and booming oilsands industry. The gas would be transported by a pipeline that would run from the delta along the entire length of the Mackenzie Valley and into northern Alberta, where it would connect with the existing pipeline network, including a new pipeline being constructed across northern Alberta to the Athabasca oilsands region (Nuttall 2006a, 2008a). The socioeconomic, political and environmental dimensions of the MGP will be explored at length in Chapter Four.

Much of the projected and anticipated activity to explore the potential of the western Arctic, including elsewhere in the Mackenzie Valley, in the Beaufort Sea and in the western High Arctic islands, is contingent on the commitment to build the MGP, and industry is hopeful the project will serve to open up this vast area, both onshore and subsequently in the Arctic offshore. As an official from Imperial Oil put it at one of the MGP community hearings in 2006, conducted as part of the regulatory review process, "….we're hopeful that, frankly, there is additional discovery of natural gas made, be it in the Mackenzie Delta or in the central Mackenzie region. But it truly is uncertain until discoveries are made."

All this potential future development will have far-reaching implications for North American economies, as well as U.S. and Canadian energy security and energy independence/interdependence. But it will also have environmental costs, as well as opening up the Arctic to tremendous economic development and exposing its mainly indigenous communities to a range of opportunities and challenges. It is only one example of what is going on across the circumpolar Arctic as, excited by the prospect that as much as 25% of the world's undiscovered reserves are to be found there, governments and energy companies are redrawing maps of the North to reflect their image of it as one of the world's last energy frontiers and a vast hydrocarbon province essential for providing future global energy supplies.

Most Arctic hydrocarbon reserves lie offshore, in the Arctic's shallow and biologically productive shelf seas, and most production activity to date involves oil onshore along the North Slope of Alaska and in western Siberia, and offshore in the Barents and Beaufort Seas. A harsh climate has made exploratory forays in search of oil and gas almost impossible, but the high costs associated with seismic surveys and drilling for hydrocarbons in remote areas have also made Arctic projects uneconomical for oil and gas companies. However, the Alaskan North Slope, the Mackenzie Delta of Canada, the Yamal Peninsula of Russia, and their adjacent offshore areas, hold enormous natural gas deposits that are projected to be developed during the next decade, while exploration for oil continues off west and east Greenland (e.g., see Nuttall and Wessendorf 2006, Rasmussen 2006).

In 2007, using north-east Greenland as a prototype for its Circum-Arctic oil and gas appraisal, the United States Geological Survey (USGS) estimated that the East Greenland Rift Basins Province could hold over 31 billion barrels of oil, gas and natural gas liquids (Gautier 2007). USGS estimates that the waters off Greenland's west coast could contain more than 110 billion barrels of oil (roughly 42% of Saudi Arabia's reserves) have already attracted interest in the territory's potential. ExxonMobil and Chevron from the U.S., Husky and Encana of Canada, the UK's Cairn Energy, and Denmark's Dong Energy are among the companies that have either already won or applied for exploration licenses from Greenland's Bureau of Minerals and Petroleum for acreage. In early January 2010, Cairn Energy announced it had obtained a second drilling rig for its exploration programme in the Sigguk Block in the Disko Bay area of west Greenland and was also turning its attention to unmapped and undrilled waters off Greenland's southern tip. In September 2010, Cairn Energy confirmed the presence of two oil types offshore Disko Island. Sir Bill Gammell, Cairn's chief executive, announced on the company's website on 21 September that "The presence of both oil and gas confirms an active, working petroleum system in the basin and is extremely encouraging at this very early stage of our exploration campaign for the Sigguk block and the entire area." Other countries, including China, have also expressed interest in being a major player in developing Greenland's energy resources (Nuttall 2008b).

In Alaska, the potential is being assessed for eventual production of heavy oil which lies in sandstone above the reservoir of the conventional light oil that has been flowing through the trans-Alaska pipeline since 1977. As production of light oil from the North Slope continues to decline, companies such as BP and ConocoPhilips see heavy oil as extending the life of the Prudhoe Bay oil fields. In the Nordic Arctic, the potential for oil and gas off Norway's Lofoten Islands and in parts of the Barents Sea shelf, such as the Hammerfest and Tromsø basins, continues to be assessed (Heitman Hansen and Midtgard 2008). The Barents Sea region is believed to contain a considerable portion of the Arctic's oil and gas and, with recent development in the Snøvhit field north of Norway and projected development by Gazprom of the vast Shtokman natural gas condensate field off north-west Russia, the area will become a major energy production and transportation hub, especially as demand for liquid natural gas (LNG) increases in the United States and Europe. And while Russia and Norway are expanding their exploration and drilling activities in the Arctic, Russia has plans to build a major energy pipeline to the strategically vital Barents Sea port of Murmansk, creating an important transportation outlet for its vast energy reserves onto the world market and transforming Murmansk into a globally important energy distribution centre. Analysts increasingly argue that not only is the Arctic going to be supplying oil and gas to meet future global demand, the region is about to become a geopolitical hub.

Russia's gas reserves amount to something like 33% of the world's known reserves and the country is the world's largest gas exporter and the third largest producer and second largest exporter of oil (Kaalhauge Nielsen 2005). Most of Russia's oil and gas production takes place in the northern parts of the country, in both its onshore and offshore regions (Moe and Wilson Rowe 2009), with Siberia producing 78% of Russia's oil and 84% of its natural gas (Weller et al. 2005). The transportation of oil and gas involves the development of infrastructure such as pipelines that cut across vast swathes of southern Russia and southern and eastern Siberia (e.g. Fondahl and Sirina 2006a, Sirina 2009), as well as along the floor of the Baltic Sea. South Korean companies have initiated feasibility studies of gas resources in the Vilyusk Basin in the Sakha Republic, including the possibility of constructing a pipeline to markets in North-East Asia, while a major gas development project with pipelines to China and eventually to other South-East Asian countries has been underway since gas was discovered near Irkutsk in 1987 (Kaalhauge Nielsen ibid.). Furthermore, large-scale development of oil and gas on Sakhalin Island off the coast of the Russian Far East continues apace. Although the oil and gas industry in northern Sakhalin has been well-established since the 1920s, new multinational multi-billion dollar offshore oil and gas projects are taking place off the north-eastern coast of Sakhalin on the Okhotsk Sea shelf and many of these follow on from the first production sharing agreements (PSAs) signed in Russia in the 1990s. These allowed multinationals to take part in oil and gas projects in the country (Roon 2006, Stammler and Wilson 2006), with the first two PSAs being the Sakhalin-1 and Sakhalin-2 projects, the latter of which was distinctive because of the involvement of international financial institutions.

Why is such expansion in Arctic energy development predicted to take place? Advances in cold regions engineering and technology certainly make exploration in the remote and harsh regions of the circumpolar North more possible, but production peaks combined with the depletion of oil and gas reserves in the world's key hydrocarbon-producing areas as well as political instability in the Middle East (where around 65% of the world's oil reserves are located) and oil-producing parts of Africa make the world's high latitude areas especially attractive for future development and for supplying much of the world's energy needs. Oil accounts for something like 40% of the world's total primary energy demand and economic conditions—and politics—are governed to a large degree by its accessibility and availability. Paul Roberts (2004: 5-6) describes our global reliance on oil and gas succinctly:

We live today in a world completely dominated by energy. It is the bedrock of our wealth, our comfort, and our largely unquestioned faith in the inexorability of progress, implicit in every act and artifact of modern existence – we need it to heat and feed ourselves, to move ourselves, to educate ourselves, to defend ourselves – everything we buy represents a measure of energy produced and then consumed.

As Roberts argues, energy has become the currency of political and economic power, influencing geopolitics and relations between states, and ensuring access to it has become the overriding imperative of the 21st century. Analysts worried that Russia is playing a game of energy politics with Europe, for instance, would doubtless agree. The economic success, but also the very survival of the energy industry, is based on being able to locate enough oil and gas to continue production to satisfy increasing global demand. In 1999, James Woolsey, a former director of the U.S. Central Intelligence Agency (CIA) and later an advisor to the George Bush administration, predicted a "peak" oil crisis, although he was not the first to do so as the idea of peak oil has influenced energy economics since M. King Hubbert first created models in 1956 to argue that oil production would peak in the United States sometime in the late 1960s. However, warnings about peak oil have taken on a new urgency with the realization that global oil supplies may be dwindling faster than energy analysts previously anticipated. Pratt (ibid.: 13) has also argued that, as far as Canadian gas supplies are concerned, "we appear to have oversold the image of an unlimited resource base".

Estimates indicate that the original recoverable oil in the earth was 2,330 billion barrels. In his influential writings on oil and gas distribution and depletion, Colin Campbell (2004, 2005) suggests that, of this amount, 90% has been discovered, 50% has been produced and that, at present, the world consumes around four barrels of its known reserves for every new barrel discovered. In terms of numbers, this equates to a total global production of something like 22 billion barrels per year, with only 6 billion barrels being discovered in that same year. The current depletion rate of the world's energy reserves has been calculated at 2.2% per year. Some of the more pessimistic energy analysts suggest that global oil supplies will peak within the next decade, although in a recent article Odell (2010) has argued against the view that the world is running out of oil. Instead, he suggests that the claims made by peak oil theorists are invalid. Claimants of the view that the world is running out of oil, he says, fail to take into account the complexity and the dynamics of the processes whereby oil reserves and production evolve, as well as ignoring the role played by economics and politics in equilibrating the markets. Whatever the arguments over the levels of supplies of oil and gas, global demand for conventional energy resources is increasing. But despite its apparent abundance and ubiquitous presence in daily life, oil is a relatively rare substance and is found only in certain geological formations. The last of the easily recoverable oil may almost be gone (Roberts ibid.) and the energy industry worries about where it can find and produce enough to meet rising demand. Although Arctic oil and gas may be found primarily in deep offshore waters, it is to these remote, difficult to reach areas that oil companies are now turning.

The prospect of a changing Arctic climate, and hence easier access to the region, is also contributing to the increased enthusiasm within industry and governments eager to pursue development there. All of this makes commentators and the media remark that global energy hunger and the disappearance of sea ice brought on by global warming are resulting in a race for resources in the Arctic (see Howard 2009 and Koivurova 2009 for a discussion of this). Oil and gas exploration and development will likely continue throughout the Arctic as climate warming contributes to reductions in sea ice, opening new sea and river routes and reducing exploration, development and transportation costs. As an indication of this trend, the United States federal Minerals Management Service announced in January 2008 that it would take bids the following month for oil and gas exploration rights and concessions in the Chukchi Sea, which separates Alaska from Siberia. The U.S. sectors of the Chukchi Sea are believed to hold some 15 billion barrels of recoverable oil and over two trillion cubic metres of natural gas.

While much of the projected oil and gas development in northern Alaska and northern Canada will take place to satisfy market demand in the United States and Canada, it is also driven by the domestic security concerns of other countries. As Chinese investment in Alberta's oilsands industry and other Canadian energy interests shows (all of which has been reported in the Canadian media as having the U.S. worried), many other countries are looking to northern Canada for their energy needs. For example, in May 2005 the Chinese-state owned Sinopec Group bought a 40%, Can$105 million, stake in Synenco Energy Inc.'s planned operation at its Northern Lights oilsands mine and secured rights to ship oil from Canada to China; also in spring 2005, the Beijing-based and also state-owned China National Offshore Oil Corporation (CNOOC) purchased 16.69% of private Canadian oil and gas company MEG Energy for Can$150 million; and PetroChina, created in 1999 as a subsidiary of the Chinese National Petroleum Company (and of which BP Amoco owns 2% of shares), has been working with Enbridge, a company concerned with energy transportation and distribution in Canada, on the proposed Gateway pipeline to take Alberta crude to the west coast of Canada for shipping to China (see Chapter Six).[16] Likewise, European countries are increasingly dependent on Russian energy sources and this influences political and economic strategies and international relations between states. Plans to build the Eastern Siberian-Pacific Ocean oil pipeline also illustrate the increasing importance of northern energy resources. Originally a project to export oil to China, interest from India, Japan and even the U.S. in western Siberian oil has broadened the potential market. In December 2009, Russia's largest shipping company announced that it would begin shipments of oil and gas eastward through the Russian Arctic's Northern Sea Route to the Pacific in summer 2010.

As a classic example of core/periphery relations, it remains a fundamental characteristic of northern resource development that most profits and benefits tend to flow from the North to southern regions. So, with the flurry of activity in Canada's North and elsewhere in the Arctic, how are indigenous communities positioned to take advantage of energy development and how will they benefit from it in the future? Will the North continue to be regarded as an extractive periphery by energy corporations, or will northern residents derive real benefits from activities that will take place on and near their lands? The rest of this book considers this question by looking at the emergence of the Arctic as an imagined hydrocarbon province and, through a discussion of plans to build pipelines and explore for further oil and gas fields, it examines a number of case studies from Canada and Alaska which illustrate some of the perspectives, interests and concerns of indigenous peoples.

CHAPTER III

TREATIES, LAND CLAIMS AND BERGER'S LEGACY

The Legacy of the Berger Inquiry

The Mackenzie Valley Gas Pipeline Inquiry was carried out during 1974-75. It was a roving commission, a process split into more formalized hearings, where expert technical and scientific testimony was presented, listened to, appraised and interrogated, and informal hearings where anyone could attend, speak and express their thoughts, views and concerns. It was notable for the flexibility and freedom Justice Thomas Berger enjoyed to travel widely and for the powers given him to compel testimony and evidence from experts and stakeholders. Berger was concerned about public participation in the process and established a fund to allow northern residents the financial means to travel to the hearings. In this way, a diversity of groups representing various stakeholders (e.g. Aboriginal and public groups) became full participants in the Mackenzie Valley Gas Pipeline Inquiry and were able to attend all hearings.

By the end of his journey, Berger had collected testimony from 300 experts on the North—including scientists, economists and oil company experts—in addition to evidence and testimony from some 1,700 northern residents. He listened to the concerns and opinions of the residents of the 35 communities situated along or near the Mackenzie River, but he also took his travelling commission of inquiry elsewhere in Canada, to cities across more southerly parts of the country because he deemed the pipeline project a national Canadian issue, not just a northern one. The hearings were broadcast on national television and radio and the transcripts of each meeting were also translated into indigenous languages and reported back to Aboriginal communities throughout the Mackenzie Valley region.

Berger concluded that oil and gas development in the Mackenzie Delta and Beaufort Sea region was inevitable and he was positive about the feasibility of developing and building an energy corridor along the Mackenzie Valley to Alberta. He believed that "in the North lies the future of Canada", reinforcing the importance of the region in the nation's geographic imaginary. However, he did not represent the North as an empty wilderness or frontier. This was an immense land inhabited by indigenous peoples with ancient cultures, he pointed out. He drew attention to the importance of the native subsistence economy, to the place of indigenous peoples within Canada's cultural life and political system, and to the testing of Canada's commitment to the environment and international co-operation.

In his 1977 report from the Royal Commission of Inquiry, *Northern Frontier, Northern Homeland*, Berger made two main recommendations to the federal government. Firstly, he was particularly concerned about the rights of Aboriginal people and the impacts of the proposed project on them. He argued that they should have some say and involvement in project development plans, but that this was unlikely given the absence of land claims and legal settlements between Aboriginal people and the government. His principal recommendation, therefore, was that a 10-year moratorium should be placed on pipeline construction until Aboriginal land claims had been negotiated, agreed upon and settled with the Canadian government. Berger was also concerned about employment, questioning whether the pipeline would provide meaningful and continuing work and career development for Aboriginal people. He was not entirely convinced by the claims of the pipeline promoters that the economic effects for the region would be positive, and he argued that large-scale projects based on non-renewable resources rarely provide long-term employment for local residents. Any employment for local people during the construction phase of the project would be unskilled, he claimed. In addition, he argued that pipeline development would erode and undermine the local economies based on hunting, fishing, and trapping and that a pipeline might actually increase economic hardship in the area.

Finally, it was Berger's conclusion that the economy of the region would not suffer undue harm if the pipeline was not constructed.

Thomas Berger's second main recommendation was a ban on construction of another proposed pipeline that would run from northern Alaska across the northern coastal plain of Yukon Territory. He feared a pipeline and energy corridor would do irreparable harm to wildlife, such as caribou, and to the people who relied on them for their livelihoods. Berger suggested the creation of a number of sanctuaries and protected areas throughout the Yukon and Northwest Territories to protect threatened species. He argued that the Porcupine caribou herd should be protected in the northern Yukon, white whales within Mackenzie Bay, and several bird species throughout the Mackenzie Valley. Berger also suggested the creation of a large reserve, a "northern Yukon wilderness park," contiguous with Alaska's Arctic National Wildlife Refuge (ANWR). The origins of Ivvavik National Park, which is located in the north-west corner of mainland Canada, on the northern tip of Yukon Territory, can be traced to Berger's recommendation.

Furthermore, on the issue of cumulative impacts, Berger believed the proposed natural gas pipeline should not be considered in isolation. He stated that construction of a gas pipeline and establishment of an energy corridor would intensify oil and gas exploration adjacent to it. He was concerned that the cumulative impact of these developments would bring immense and irreversible changes to the Mackenzie Valley and the entire Western Arctic. Berger's eloquent articulation of the North being a "homeland and a heritage that we are called upon to preserve for all Canadians" resonated with politicians and the public alike. Although Berger noted that the *Expanded Guidelines for Northern Pipelines* tabled in Canada's House of Commons on 28 June 1972 called for an examination of proposed pipelines from the point of view of cumulative impact, the issue of cumulative impact has not been specifically addressed to date, nor was it an explicit concern for the boards charged with overseeing the public hearings of the Mackenzie Gas Project in 2006-07 (see Chapter Four).

The Berger Inquiry, as it became known (very few refer to it, nor indeed remember it, as the Royal Commission of Inquiry, so synonymous is it with the man who led it), was significant in that it made an event out of public hearings for the assessment of environmental impacts. No major frontier resource development project in Canada had ever before been reviewed through public participation before construction was permitted (Nassichuk 1987). Only a handful of environmental impact studies had ever been conducted in the Mackenzie Delta, perhaps the most significant being *The Environmental Impact of the Proposed Mackenzie Delta Gas Development System* carried out by Gulf Oil Canada, Imperial Oil and Shell Canada in 1976; and only a few more major environmental impact studies have been carried out since (e.g. *Pipeline Environmental Effects* by Polar Gas in 1984; the *Environmental Assessment of the Fort Liard Gas Pipeline and Facilities* by Chevron Canada Resources in 1999; and the various studies needed for the Mackenzie Gas Project).

The Berger Inquiry also played a significant part in a decade that some commentators argued "thrust the North into the Canadian consciousness" (Dacks 1981: 1), and "provided the sharpest focus for the political issues of the 1970's. The list of components in this debate is long and curiously disparate, and includes environmental protection, native rights, economic nationalism, energy conservation, the limits of high technology, political sovereignty, public participation, and government regulation" (Page 1986: x). It is remembered for its accessibility, for its emphasis on accountability, and for the sense that it was a journey of discovery into the lands and lives of northern Canada's indigenous peoples.

For most Canadians at the time, the North was a remote part of the country few had much knowledge about, even if the idea of North was essential to Canadian identity and nation-building.[17] The images of the hearings and of Berger himself conducting his inquiry remain both iconic and symbolic, particularly when contrasted with the public hearings process for the Mackenzie Gas Project 30 years later, as I shall discuss in Chapter Four. Berger huddled with local people in schoolrooms and community halls, hearing their stories of life on the land, pouring over maps of places of local cultural, historical, economic and spiritual significance. Listening to people talk of the land and the animals that sustained them, he tried to grasp the significance of indigenous knowledge, the cultural power of human-environment relations, and the actuality of the land. For many who participated in the hearings, the memories of this dedicated and sympathetic engagement remain strong. Reflecting on his report 12 years after its publication, Berger remarked that, "I urged in 1977 that we think of development not in terms of large-scale, capital-intensive frontier projects, but in terms of strengthening the traditional economy on which indigenous peoples throughout the Arctic and Subarctic have depended for hundreds and hundreds of years" (Berger 1989: 37).

The Berger Inquiry was a mapping of Canadian cultural space. The meetings were opportunities for local people to tell stories, and each hearing was regarded as a forum to continue a series of conversations about human-environment relations and the future of Canada and its North. It continues to provide fertile ground for Canadian literature and academic research. For example, Elizabeth Hay's 2007 novel *Late Nights on Air*, which focuses on the personalities working at a Yellowknife radio station in 1975, is set against the backdrop of the inquiry and the local tensions arising from Aboriginal political activism. "Tom Berger," writes Hay, "was a pearl. He listened with grave, courteous, uncommon openness, being a careful speaker himself, and one who took his time" (Hay 2007: 84). Hay continues to dwell on "Berger's distinctive voice" because it was something that characterised the uniqueness of the event. It was a voice, she writes, that

became familiar to everyone in the North. Those who heard it would recognize it immediately, no matter how many years had gone by, his firm, thoughtful, soft-spoken voice soliciting expert testimony about the social, environmental, and economic impact of the pipeline, but also the opinions of anyone who would be affected by what would be the biggest project ever built in the history of free enterprise, if it went ahead.

(Hay ibid. 85)

The inquiry remains an important northern trope and continues to provide inspiration for the way things should be done, but it also provides a starting point for the analysis of the history of contemporary northern Canada, its industrial transformation and its place within the nation. It has resonance today because, as Hay articulates eloquently in her work of fiction,

At stake was something immense, all the forms of life that lay in the path of a natural gas pipeline corridor that would rip open the Arctic, according to critics, like a razor slashing the face of the Mona Lisa.

(Hay ibid.: 83)

Berger found critical gaps in information about the northern environment, environmental impacts, and engineering design and construction on permafrost terrain and under extreme Arctic conditions. He called for a continuing process of northern science and research which would provide an independent body of knowledge. It would not be overstating things to say that Thomas Berger changed the way Canadians view resource development. He highlighted and humanized a complex political, cultural and environmental issue, and his report also pointed out that his inquiry was about more than pipelines; it was about protecting the northern environment and the future of northern peoples. The Government of Canada listened to Berger's recommendations and placed a 10-year moratorium on the issuance of exploration rights for oil and natural gas in the Mackenzie Valley and southern NWT. Yet at the same time, a decline in oil and gas prices also meant that energy companies felt less favourable towards investing in northern projects.

The Mackenzie Valley Pipeline Inquiry set an international standard for critical and cross-cultural assessment of proposed development plans and options. Despite its wide acclaim, however, the model was never again used in Canada, although it influenced subsequent deliberations and practice in environmental assessment, particularly the federal Environmental Assessment and Review Process (Gibson 2002). The Berger Inquiry provided space for the expression of native concerns and provided an opportunity for the formation, the shaping and the articulation of arguments for indigenous land claim settlements and greater self-determination. The inquiry coincided with the 1975 signing by the Governments of Canada and Quebec and the Inuit and Cree of northern Quebec of the James Bay and Northern Quebec Agreement, which included the first of many claims-based environmental assessments (Gibson ibid.).

Abele (1983) and Coates and Powell (1989) argue that the prominence and resonance of the Berger Inquiry may be explained by the role it played in the transformation of the fundamental social relations of Aboriginal societies in the Mackenzie Valley, as well as the rejection of colonialism and a realignment of relations between North and South in Canada. The inquiry itself was a crucial phase in the history of the social, cultural, economic and political transformation of northern Canada which began with prolonged contact with European capitalism, most notably through the fur trade in the 18th century. Although outside influences shaped social and economic changes in the Mackenzie Valley, Abele argues that these influences did not decisively draw Aboriginal peoples such as the Dene into capitalist society. However, economic development and the rush for energy resources in Canada following the discovery of oil at Prudhoe Bay in Alaska in the late 1960s threatened to sever Aboriginal peoples' relationships with the land. The Berger Inquiry coincided with attempts by Aboriginal people to communicate their anxieties and argue for their rights, as well as the politicization of native culture and the emergence of native politics, illustrated by the formation of organizations like the Committee for Original Peoples' Entitlement (COPE), which had been established in 1970 to represent the Inuvialuit of Canada's western Arctic. In 1976, COPE assumed the responsibility of negotiating a land claim with the Canadian government. The claim included demands for ownership and control of land and natural resources, as well as the importance of recognizing indigenous voices in the future development of the Inuvialuit region.

The title of Berger's report, *Northern Frontier, Northern Homeland*, emphasized that the Mackenzie Valley and Mackenzie Delta was the social, cultural and spiritual home of a large number of Aboriginal people (Inuvialuit, Gwich'in and Dene), in addition to being an area of increasing economic importance to industry keen to satisfy energy demand in southern markets. Development, argued Berger, needs to conform to the wishes of those who live there. More starkly, Berger predicted that the "social consequences of the pipeline will not only be serious—they will be devastating". As he emphasized:

I discovered that people in the North have strong feelings about the pipeline and large-scale frontier development. I listened to a brief by northern businessmen in Yellowknife who favour a pipeline through the North. Later, in a native village far away, I heard virtually the whole community express vehement opposition to such a pipeline. Both were talking about the same pipeline; both were talking about the same region - but for one group it is a frontier, for the other a homeland.

The report reinforced this view by describing "White" and "Native" worldviews and attitudes to the land. Writing about the testimony of Dene witnesses, for example, Chambers (1989) has shown persuasively how speakers employed metaphor, irony and personal stories to call into question the morality of white people and the social and institutional practices of Euro-Canadian society that had altered, and were continuing to transform, the northern landscape and Dene lives and livelihoods. Dene used the hearings to tell stories about Dene ideas of respect—being respectful to, and of, the other, including all life, both human and non-human, as well as the land itself. Patrick Scott (2007: 3) writes that, while it is not surprising that Berger gets most of the credit for the success of the inquiry, the reason for the "real success rests in the storytelling, not in the man charged with listening to the stories". The stories told by northern indigenous people, as Scott reminds us, were not only traditional ones, or stories told about special places: "mostly we heard the stories of the people's day to day lives and what they valued."

At the same time, through their testimony and by telling stories, Aboriginal people who spoke at the hearings called for people to imagine themselves into the future, and to envision it through children who were yet to be born. Scott (ibid.: 3) calls the transcripts from the hearings "a marvelous collection that, at the time of telling revitalized Dene culture". Villebrun (2002) also examined the public speaking experiences of Dene from the Deh Cho region of the central Mackenzie Valley, who voiced their understanding of public conduct and knowledge sharing as being distinct from a Western way of public performance. Villebrun argues that we can understand Dene testimony as a narrative on colonization and a challenge to established epistemological assumptions about northern development and the land.

Despite the high public profile and political impact of the Mackenzie Valley Pipeline Inquiry, Berger's report has its critics, who continue to debate the process and its impacts. One criticism is levelled at the inquiry's impression that there was overwhelming Aboriginal opposition to the pipeline. Examining the Inuvialuit participation in the hearings, Cargill (2002) suggests that Inuvialuit who spoke expressed a wide range of opinion, including a positive interest in large-scale development in the Mackenzie Delta. Cargill argues that, although Inuvialuit culture and the northern economy at the time rested on a strong relationship with the land, this did not necessarily mean that the Inuvialuit were against oil and gas development. She shows how the subtleties of the Inuvialuit position prepared the way for a new and integrative approach to resource development in northern Canada. Rather than being an exercise in coercion by southern culture, the Berger Inquiry was a process that allowed for the participation of Aboriginal people on their own terms, even if some of their pro-development perspectives were not necessarily highlighted to a great extent in the final report. Stabler (1977) also offered a socio-economic critique of the Inquiry, arguing that, throughout the report, Berger emphasizes the importance of fur, fish and game to the native economy, and that wage employment in Mackenzie Valley communities serves to reinforce the native economy and the native culture. Stabler points out that the report does not attempt to provide a systematic economic evaluation of the returns from hunting, fishing and trapping.

Campbell (1985) identified two areas of debate that emphasized scientific uncertainty during the hearings. Firstly, experts debated the adequacy of knowledge about the impact of pipeline development. Industry proponents were required to demonstrate that their plans for the project would have limited negative environmental effects, and the scientific experts they mobilized to support their position on the project cited scientific studies suggesting that negative effects would be minimal. Critics of the project, on the other hand, did not need to muster an equal amount of evidence. Instead, they simply claimed that the defenders' knowledge was inadequate because insufficient research had been conducted to demonstrate the environmental integrity of the project. They put forward a different interpretation of the evidence submitted by the defenders, argues Campbell, without necessarily claiming the data were wrong. Secondly, Campbell suggests that while the critics claimed uncertainty and the defenders claimed certainty in scientific knowledge, the defenders nonetheless argued that uncertainty was an entirely normal and reasonable feature of science and that scientific uncertainty in itself was not enough to put a halt to pipeline development. The defenders of the Mackenzie Valley project argued that scientific uncertainty was manageable, with critics arguing the contrary. Nelkin (1975) suggested that critics of major projects need not necessarily offer equal evidence to offset the expertise of the proponents but that they merely need to be good at pulling apart, or deconstructing, positions advanced by the proponents. Campbell's argument is that the Berger Inquiry demonstrated that claims of scientific uncertainty can have the same authoritative force as claims of certain knowledge which is offered and represented as authoritative.

Almost concurrent with the Berger Inquiry, Kenneth M. Lysyk headed the Alaska Highway Pipeline Inquiry, and reached many of the same conclusions as his counterpart did for the Mackenzie Valley pipeline proposal. In August and September 1976, the Foothills (Yukon) Group of companies had submitted proposals for construction of the Foothills (Yukon) Project (Cowling 2001). The project was more commonly referred to as the Alaska Natural Gas Transportation System (ANGTS) and the plan was for a pipeline to transport gas reserves from Alaska's North Slope to American consumers. The pipeline was planned to follow much of the route of the recently constructed Alaska oil pipeline, then cross the Alaska-Yukon border and continue through Yukon, northern British Columbia and Alberta to the Canada-United States border in southern Saskatchewan. Lysyk concluded that most of the economic benefits of the pipeline, as well as its profits, would end up outside of the Yukon, but that there was large conditional support for the project if the land claims of Yukon First Nations were settled and the pipeline companies agreed to a financial agreement to mitigate negative social impacts of the project. The Alaska Highway Pipeline did not pose particular environmental concerns, as the right of way was already in place in the form of the Alaska Highway corridor. Lysyk also suggested that an inter-jurisdictional planning and regulatory agency was needed to oversee pipeline construction, and finally reported that the pipeline should be delayed for four years while the remaining land claims were settled. The Lysyk Inquiry will be discussed in more detail in Chapter Five.

Current Oil and Gas Activity in Northern Canada and Traditional Knowledge

The Berger Report and the federal government's subsequent acceptance in 1977 of Thomas Berger's recommendation for a 10-year moratorium was something of a setback to the industry, but oil and gas exploration and production nonetheless continued elsewhere in the Canadian Arctic. Drilling activities also took place off the coast of Labrador. These continuing prospecting and development activities were explained partly by the 1980 National Energy Program (NEP), a bill which had been introduced by the Liberal government of the time. The NEP was somewhat controversial and had the effect of making Canadian exploration less attractive to those parts of the oil and gas industry that found the new jurisdictional environment not to their liking. However, the NEP aimed to reduce Canadian dependence on foreign oil and to reduce foreign ownership of the Canadian oil industry (any venture had to be at least 50% Canadian-owned before going into production), and it allowed companies to write off more than 100% of the costs incurred for exploration and development in the North and remote areas, while not making such concessions for the same activities in southern Canadian provinces. It also provided for a 25% federal government share in all oil and gas discoveries in the North and offshore, for which Ottawa had to pay nothing.

Following the Berger Inquiry, energy companies kept their pipeline dreams alive and as long as world oil prices rose it was only a matter of time to wait for the right political and economic environment. However, with gas and oil prices plummeting in the early 1980s, the incentives for building a pipeline through the Mackenzie Valley or along the Alaska Highway soon dissipated. But it was not long before renewed interest on the part of Gulf Canada, Shell Canada and Esso Resources Canada (Imperial Oil) led to Mackenzie Delta gas export hearings being held by the National Energy Board in 1989, with Foothills Pipe Lines also filing an application for approval to construct a Mackenzie Valley pipeline. At the time, Berger (1989: 38-9) commented that "the conditions that I felt had to be met before a pipeline could be considered have, to a large extent, been met. It is logical that the industry should renew proposals to build a Mackenzie Valley pipeline, and I am content to leave it to the National Energy Board, to northerners, to the governments of the Northwest Territories and Yukon and last, but not least, to the government of Canada to make their own choices about such matters."

Exploration and drilling continued and, by the mid-1990s, there were almost 2,000 wells drilled north of 60° latitude. Furthermore, negotiations for comprehensive land claims between the Canadian federal government and Aboriginal peoples had begun to reach conclusions and final settlements. The Inuvialuit settled the first comprehensive land claim in Canada's far north in 1984, and this was followed by settlements with the Sahtu Dene and Gwich'in (this will be discussed in further detail later in this chapter). New exploration licenses were issued for the southern Northwest Territories and the central Mackenzie Valley in 1995, signalling a return to significant oil and gas exploration and investment in Canada's North. By the late 1990s, interest in building a pipeline along the Mackenzie Valley was again increasing. In 2000, a sale of mineral claim rights proved lucrative for the Canadian government and energy policy in the United States began to favour the stepping up of oil and gas production from North American sources. In 1999 and 2000, companies acquired exploration rights to lands across much of the Mackenzie Delta and adjacent offshore areas in the Beaufort Sea.

As conventional crude and natural gas production begins to decline in the traditional producing areas of Canada's western provinces, and while unresolved land claims in the southern Mackenzie Valley affect the issuance of licenses (although new gas discoveries are being made there), oil and gas companies are continuing to turn their attention to more northerly regions. Interest has again been rekindled in examining what it would take to develop gas from the Mackenzie Delta and Beaufort Sea. The gas grid is being pushed northward, especially as concerns rise that natural gas supplies south of the 60th parallel may be inadequate to meet future demand, particularly from the United States. The strength of western Canadian gas prices has also been viewed by industry as a reason for developing new sources of supply. From the perspective of energy companies, these and other related factors warrant a new strategic and commercial review of northern Canadian oil and gas development.

Currently, the focus is on three main areas, the Sahtu region of the central Mackenzie Valley, especially in the Colville Hills, the Mackenzie Delta and the Beaufort Sea. Although the search for hydrocarbons has been extensive over the last few decades, only seven oil and gas fields are in production in the Northwest Territories—four gas fields and one oil and gas field in the southern Northwest Territories; the Norman Wells oil field in the central Mackenzie Valley; and the Ikhil gas field in the Mackenzie Delta. There are no fields producing in Nunavut or in offshore Arctic waters (INAC 2005). Investment in new exploration of northern oil and gas reserves has been, to a large extent, largely dependent on decisions concerning the Mackenzie Gas Project, which energy companies hope will go ahead as it will open up possibilities to develop and produce new fields. The Mackenzie Gas Project (MGP) will be based initially on production from three gas anchor fields in the Mackenzie Delta but, as I explore in Chapter Four, industry has long considered that it will open the way to investment in further oil and gas exploration onshore and, eventually, in the Arctic offshore.

In Canada, the management of petroleum resources on Crown lands north of 60° N (except in Yukon) is exercised under federal legislation. The Canada Petroleum Resources Act and its regulations govern the granting and administration of Crown exploration and production rights and set the royalty regime. The Canada Oil and Gas Operations Act governs the regulation of petroleum operations and the requirements for associated benefits. Land, royalty and benefit matters are managed by the Northern Oil and Gas Branch of Indian and Northern Affairs Canada (INAC) on behalf of the Minister of Indian Affairs and Northern Development. The National Energy Board (NEB), an independent federal agency, takes the lead role in the regulatory approval of operations.

The process of a Call for Nominations and Bids enables industry to specify blocks of land of interest for oil and gas exploration. The Northern Oil and Gas Branch then embarks on a process of consultation with Aboriginal groups and communities to ensure their views have some legitimacy and influence over the decisions eventually made for the issuance of rights and licenses. This consultation process gives communities an opportunity to discuss concerns over what exploration and development would mean for areas of environmental sensitivity, as well as places of cultural or spiritual importance. In recent years, the Government of Canada and the Northwest Territories government have approved the Protected Areas Strategy (PAS) for the Northwest Territories, which aims to conserve biological diversity and natural and cultural resources in the Northwest Territories. During the consultations which are held with Aboriginal groups on rights issuance, a review is also completed to ensure that lands that may have been nominated through the PAS process are not included in the calls. INAC also consults with other federal departments, territorial governments and agencies. Environmental considerations play an important role when issuing permits to energy companies for land and water use and other work permits. The terms and conditions of a Call for Nominations and Bids reflect the results of this consultation process.

For Aboriginal communities, traditional knowledge studies are essential for the success of this consultation process. As used in Canada, "traditional knowledge" is a term that describes a body of knowledge generally held by indigenous peoples about their cultural and physical landscapes. A debate on the use of the term "traditional" has long characterized some academic and professional discourse; however, the term "traditional knowledge" has become an established phrase, used by indigenous peoples in environmental impact assessment processes and acting as a form of cultural shorthand to embody a diversity of cultural values, attitudes and ways of knowing. In the Northwest Territories, for example, the Gwich'in Tribal Council, has its own Gwich'in Traditional Knowledge Policy which describes and defines Gwich'in traditional knowledge, and sets out how it will be collected and used. Gwich'in Traditional Knowledge is defined by the Gwich'in Tribal Council as

> *that body of knowledge, values, beliefs and practices passed from one generation to another by oral means or through learned experience, observation and spiritual teachings, and pertains to the identity, culture and heritage of the Gwich'in. This body of knowledge reflects many millennia of living on the land. It is a system of classification, a set of empirical observations about the local environment and a system of self-management that governs the use of resources and defines the relationship of living beings with one another and with their environment.*

Traditional knowledge can make a substantial contribution to the environmental impact assessment process and indigenous peoples worldwide have argued the importance of including it in any discussion of the environmental impacts of resource development (Nuttall 1998). As an example of this, the Gwich'in Traditional Knowledge Study of the Mackenzie Gas Project Area was initiated in response to regulatory requirements under various acts and policies for the inclusion of traditional knowledge in environmental impact statements. A large amount of traditional knowledge information was gathered through interviews with community members by the Gwich'in Social and Cultural Institute (GSCI). The collection and description of traditional knowledge in the Gwich'in Settlement Region was assigned to the GSCI by the Gwich'in Tribal Council.[18]

The Gwich'in Traditional Knowledge Study of the Mackenzie Gas Project Area was initiated by the proponents of the Mackenzie Gas Project, funded by Imperial Oil and coordinated through the Mackenzie Project Environmental Group (MPEG). The MPEG consists of consultancy firms AMEC Earth and Environmental, Kavik-AXYS Environmental Ltd, Tera Environmental, and Golder Associates Ltd. From the regulatory perspective, the purpose of this study was to meet the specific traditional knowledge information needs of the Mackenzie Gas Project, facilitate meaningful community participation in the environmental and socio-economic impact assessment process, and ensure compliance with all regulatory requirements for using traditional knowledge in an environmental assessment, including those in the Mackenzie Valley Resource Management Act, the Canadian Environmental Assessment Act, the National Energy Board Act, and land claim agreements such as the Inuvialuit Final Agreement, the Gwich'in Land Claim Settlement Act, and the Sahtu Dene and Métis Land Claim Settlement Act. Of concern to the GSCI was the need to ensure the protection of important cultural and natural areas and species, especially on or near the route of the proposed pipeline, and for the project proponents to recognize that such protection is necessary for the continuation of Gwich'in cultural practices and traditional livelihoods. A number of reports have been produced by the GSCI and other Aboriginal organizations that aim to illustrate the cultural, spiritual and economic importance of the land as revealed through community testimony. During traditional knowledge workshops, for example, elders were recorded talking about important sites for hunting, fishing or berry picking, or sites of cultural heritage, that are directly on the proposed pipeline route. For energy companies there may be nothing immediately significant about such areas, but pipeline construction would sever a vital link between people and the land.

Energy Development and Land Claims

In the new millennium, oil and gas exploration and development in northern Canada take place in a radically different social, political and economic context from during the time when Thomas Berger's inquiry into the Mackenzie Valley Gas Pipeline project travelled through the North and other parts of Canada and held its hearings. Since Berger made his recommendations in the late 1970s, perhaps the most significant change in the Canadian North has been the settlement of comprehensive land claims with several Aboriginal peoples. These have followed on and resulted from protracted and complex negotiation processes concerning land use and resource ownership. Canada's Northwest Territories has a total population of some 42,000, and Aboriginal people—mainly Inuvialuit, Dene and Métis—comprise approximately half of this figure. In Nunavut, 82% of the total population of 27,000 is Inuit. In Yukon Territory, some 23% of the 31,000 people living there are Aboriginal.

For many Aboriginal communities, hunting, trapping, fishing and gathering remain important social, cultural and economic activities. They provide traditional foods from the land, lakes, rivers and coastal waters, and their consumption is a fundamental part of social identity and personal and cultural well-being, in addition to being a celebration of community and of the intricate relations between humans and animals (Nuttall et al. 2005). In the modern North, however, although traditional hunting, fishing and trapping practices remain vital to the daily lives of Aboriginal people, commercial fishing, diamond mining and the oil and gas industries increasingly provide, or promise to provide, employment. This is currently so for the Northwest Territories, and the energy and mining industries may become the main driver of the Nunavut and Yukon territorial economies.

Across northern Canada there is a flurry of exploration activity and a rush to stake claims to the land's riches. Oil and gas exploration seismic survey crews are busy producing "CAT scans" of the Earth's subsurface, while prospectors are searching out and staking claims to commercially viable deposits of valuable minerals. The recognition of indigenous peoples' rights and settlement of land claims has meant that Aboriginal communities have a clear understanding of legal entitlement to energy and mineral resources. They do much to exercise this right and many have entered into resource development projects through joint ventures with industry and government, impact benefit agreements and environmental monitoring projects. Aboriginal peoples have long been part of a world system through fur trade economies and, more recently, through other kinds of enterprise such as commercial fishing. Today, oil, gas and minerals are part of their cognized environment (Rappaport 1968) and, while some communities are divided over whether energy development and mining are industries indigenous peoples want on their lands, most are in agreement that, if more kimberlite exploration is to take place, and if oil and gas wells are to be sunk, then indigenous peoples have the right to benefit from the development of these and other extractive industries. In Nunavut, for example, there is territorial government support for oil and gas exploration in the territory's sedimentary basins and for opening Nunavut up to considerable activity on an annual basis. Under the Nunavut Land Claims Agreement, Inuit have a major role in ensuring that resource development must be done in a way that respects the environment and ecosystem integrity, but that it must also be done in a way that respects the values of Inuit and provides long-term economic benefits for Nunavut.

Over the last 25 years, land claims and self-government negotiations between the Government of Canada and Inuit and First Nations have resulted in the recognition of Aboriginal rights. The settlement of what are known as comprehensive land claims has radically altered the map of the Canadian North where there were outstanding claims to Aboriginal title over traditional lands. In most cases, these have transformed the political, social, economic lives of indigenous peoples because, as Bone (2003: 192) has observed, "They provide the means and structure to participate in Canada's economy and society and yet retain a presence in their traditional economy and society."

Land claims processes—in effect, modern treaties—are unique to North America in the northern circumpolar regions and, in Canada at least, are a first step towards self-government. As Irlbacher-

Fox (2005: 1152) states, the negotiating processes are "compelled by legally recognized aboriginal rights to lands and resources". The Inuvialuit of the western Arctic reached a land claim in 1984, the Gwich'in did so in 1992, the Sahtu Dene in 1994, and the Inuit of Canada's Eastern Arctic were given their own self-governing territory, Nunavut, in 1999. In the Northwest Territories, the Tlicho (Dogrib) First Nation signed a land claims agreement with the Canadian government in 2003, negotiations for land, resource and self-government rights continue with the Dehcho First Nations and the Akaitcho Dene while negotiations for self-government are in progress with the Inuvialuit, Gwich'in, and the Sahtu Dene community of Deline. In Yukon, the Umbrella Final Agreement, which was reached in 1988 and finalized in 1990, is used as the framework or template for individual land claims agreements with each of the 14 Yukon First Nations. Since then, ten First Nations have signed and ratified an agreement; another two have signed agreements which were not ratified after being defeated in referendums; and two are still being negotiated. In most parts of the provincial North, i.e. the northern areas of Alberta, British Columbia, Saskatchewan, Manitoba and Ontario, treaties signed in the late 19th and early 20th centuries remain in effect. The Inuit and Cree of northern Quebec signed the first modern treaty, the James Bay and Northern Quebec Agreement, with the Canadian government in 1975, while the Inuit of Labrador signed the Labrador Inuit Land Claims Agreement in 2005.

The 1984 Inuvialuit Final Agreement extinguished Inuvialuit rights and interests in land in exchange for ownership of 91,000 km² of land, cash compensation of Can$170 million, preferential hunting rights, participation in resource management, subsurface mineral rights to a small area of land, and a provision for future self-government. In 1992 the Gwich'in Comprehensive Claim Agreement gave the Gwich'in ownership of 22,331 km² of traditional lands with subsurface mineral rights to one-third of that area. Other rights and benefits include Can$75 million, a share of Mackenzie Valley resource royalties, participation in the planning and management of land, water and resource use, and a federal commitment to negotiate self-government. The Sahtu Dene and

Métis Agreement (1994) gave beneficiaries 41,437 km² of land, with subsurface rights over 1813 km², and it contains similar provisions to the Gwich'in agreement. The Nunavut Final Agreement of 1993 (which preceded the Nunavut Act establishing the government of the new Nunavut Territory) established fee simple title for approximately 21,000 Inuit beneficiaries to just over 18% of the total territory of Nunavut, which includes mineral rights to over 36,000 km².

On paper at least, these agreements have meant three significant things for Aboriginal peoples in northern Canada. Firstly, by providing cash compensation and setting resource royalty levels, they have established administrative structures and provided the financial resources to make it possible for Aboriginal communities to survive and function effectively in the mainstream Canadian economy. Secondly, they have defined use of lands and resources, and included guarantees to Aboriginal peoples for specific access to natural resources, including subsurface minerals. Thirdly, they have initiated co-management regimes, which are forms of shared governance, for decision-making over natural resources, land-use planning, wildlife management and environmental issues (Bone 2009). Unlike earlier agreements and treaties, which emphasized the exchange of lands for various forms of compensation, comprehensive land claims agreements in Canada's North have emphasized instead the importance of land and resource governance over land sales (Irlbacher-Fox ibid. 1152).

With comprehensive land claims, Aboriginal rights have been more clearly defined. Significantly, as was the case in Alaska with the Alaska Native Claims Settlement Act of 1971, the negotiation over land claims and rights over use and access to resources occurred in the face of plans for megaproject development. Irlbacher-Fox (ibid. 1152) observes correctly that, while they have been misunderstood as taking an anti-development stance, "one of the basic goals of indigenous peoples has been to participate in and control aspects of development—to engage in it rather than be excluded from it". Land claims are symbolic for indigenous peoples in the ways they acknowledge recognition of their rights as well as provide them with a means to ensure economic, social and cultural survival.

Energy and mining companies are often powerful transnational players with experience of working in many regions of the globe and their economic power and legal strength can often be overwhelming to a small Aboriginal community that has little opportunity to engage in discussion over the impacts and benefits of resource development. Where Arctic residents have opportunities to capture some of the economic benefits from industrial development, both through employment and corporate investments, benefits in the form of improved public infrastructure, educational services and health care can be significant (e.g., as has happened in Alaska's North Slope Borough). Yet the large-unit size and sheer scale of most oil and gas-related development can actually increase the dependence of local communities and regions upon national governments and transnational corporations. The financial and employment benefits that may flow to local communities as a result of oil and gas development may be countered by increasing dependence on national government for the provision of infrastructure, environmental assessments, anti-pollution measures, occupational health and safety policy, and for policy responses to the uncertainties and fluctuations inherent in the global energy economy.

Land claims as well as other legislation, however, have meant that Aboriginal title to lands and resources and the duty to consult must be recognised by anyone wishing to do business in Canada's North. In Yukon Territory, the Yukon Oil and Gas Act gives First Nations the right to participate in any decisions related to oil and gas development. Despite the absence of a Final Agreement over land claims, First Nation consent is required before any third party oil and gas interests can be created. Under the act, there is an additional requirement for impact benefit agreements. This and legislation in other areas means that, throughout Canada's North, companies must enter into agreements and negotiate impacts and benefits agreements with Aboriginal communities. In the Northwest Territories, Aboriginal involvement in decision-making for resource development is also guaranteed under the Mackenzie Valley Resource Management Act, which grew out of the comprehensive land claims agreements of the Gwich'in, Sahtu Dene and Métis. The Government of Canada, through the National Energy

Board and Indian and Northern Affairs Canada (INAC) controls more than 90% of the petroleum subsurface rights in the NWT. However, those Aboriginal groups that have concluded land claims agreements have responsibility for subsurface rights and royalty regimes in parts of their territories. Senior executives of Imperial Oil, for example, have stated publicly that the Mackenzie Valley gas pipeline will never be constructed without the support of Aboriginal communities in the North.

Treaty-Making and Canada's Great Resource Storehouse

The fundamental basis for comprehensive land claims and access and benefits agreements resulting from resource development lies in the British justice system and in the continuing cultural and legal power of Aboriginal title. In 1763, at the end of the Seven Years' War between the British and French, Great Britain's King George III issued a Royal Proclamation claiming sovereignty over North American territory previously claimed and occupied by the French. He declared that territories and hunting lands west of rivers draining into the Atlantic should remain lands for "Indian Nations" and the Royal Proclamation set out a broad policy for how the British Crown could obtain rights to Indian lands. Treaties and military alliances with Indians had been commonplace in British North America and the Royal Proclamation made clear that only the Crown could purchase land from Indians on a nation-to-nation basis at public meetings convened for that purpose. This provision prevented land sales to private individuals or land purchases by other governments (Irlbacher-Fox ibid.).

When Canada was formed as a nation in 1867, Ottawa continued the treaty-making tradition. That same year, the new federal government negotiated with the Hudson's Bay Company for the purchase of Rupert's Land and the vast North-West Territories, the lands that now make up the majority of the western and northern parts of the country. With dreams of nation-building and a united British North America, the federal government and the political and economic elite of the new Dominion of Canada "had their eyes on the prize—the vast lands and resources of Canada's North-West" (Calliou 2006: 304). The national dream of breaking down the frontier, developing its resources and settling the North-West quickly became a matter of policy, consolidated by Prime Minister John A. MacDonald's National Policy of 1879, which sought to attract pioneering immigrants from Europe to clear the lands of the west, settle there and farm. The lure for anyone wishing to immigrate to Canada was a free homestead. But first, dealing with outstanding land claims and treaty-making with the Aboriginal peoples were matters of urgency if expansion was to proceed and settlement to follow.

Between 1870 and 1921, Treaty parties and Halfbreed Scrip Commissions (to deal with Métis scrip), as Canadian representatives of the British Crown, ventured into indigenous lands and negotiated and signed 11 numbered treaties with the Aboriginal peoples of north-west Canada, with the purpose of obtaining land for settlement and allowing access for resource exploitation. This opened up enormous areas of fertile land between the Canadian-United States borderlands, extending through the resource-rich Rocky Mountains and west to the Pacific coast, for settlers and resource developers, as well as for the building of railroads. Indigenous peoples' land and resource rights were not addressed by treaties in other parts of northern Canada (i.e. in Labrador, Yukon, and what is now Nunavut and Nunavik).

Of relevance for themes discussed in this book, Treaty 6 was signed in 1876 and covered the areas now a part of central Alberta and Manitoba; Treaty 8 was signed in 1899 with the Aboriginal people of northern Alberta and north-east British Columbia; and Treaty 11 was signed in 1921 for much of the Mackenzie District of what is now the western part of the Northwest Territories (between them, Treaties 8 and 11 covered the Mackenzie Basin area). Treaty 11 was the last of the numbered treaties. Its significance, however, lies in the fact that oil had been discovered at Norman Wells. Because the land in the area had long been deemed inadequate for settlement and agricultural development, the federal government in Ottawa had been reluctant to engage in a treaty-making process with the indigenous population. Oil, and getting it out of the ground, moved the government to initiate treaty negotiations and this says much about how Canada has viewed the reasons for making treaties with Aboriginal peoples.

Historians of the Canadian North-West have demonstrated that Aboriginal peoples sought at various times to initiate treaty discussions, particularly in the northern areas covering present-day Alberta, Saskatechewan and Manitoba. These areas, however, were outside the fertile agricultural districts and the lands the government wished to encourage immigrants to settle in. Negotiating and concluding treaties had no immediate relevance or significance for a succession of governments which viewed the only reason for treaty-making to be one of obtaining lands and making way for settlement and resource extraction. Revisionist historians have argued that, rather than being unwilling recipients of treaties which were imposed upon them by the government of Canada and its representatives, indigenous peoples were active agents in treaty negotiations, making their demands clear and refusing to accept less than modest agreements. Sharon Venne, a Cree writer and expert on treaty rights, argues that treaty-making was part of Aboriginal culture and well-established by the time the agents of the Crown entered the territories of indigenous peoples in north-west Canada. For the Cree of what is now northern Alberta she writes,

> You only have to go back a short way in the history of our Cree peoples, who made treaties with our neighbouring Indigenous nations. There were wars between the nations so there was a need for peace treaties. Peace treaties are known to the Cree. The Cree made a peace treaty with the Dene that is still in place. The Cree-Dene Treaty—concluded before the coming of the non-indigenous peoples—was to demarcate our territories. The demarcation is known as Peace River: north of the Peace River is Dene land, and south of it is Cree territory. When I cross the Peace River going north into Dene territory, I always give thanks to the Cree for letting me come into their territory.

(Venne 2006: 2)

The federal government knew that treaties were essential to the opening up of the west and the vast potential of the wheat-producing prairies. As Calliou (ibid.) writes, however, Aboriginal peoples were not insignificant actors in the historical development of the Canadian North-West. At the same time as the expansionists were acting out the national dream of opening up the frontier, indigenous groups, who had long been strong trade partners with non-indigenous people and an active part of the capitalist economy of Canada, were in the midst of experiencing profound changes to their lives and lands. Buffalo herds were in decline, and disease and famine were widespread. They recognized the inevitability of both further social and economic change and the westward movement of many thousands of settlers. This westward expansion was beginning to reach into the North. Writing about his travels in 1899 in the Athabasca and Peace River districts of what is now northern Alberta, Charles Mair described this change, which was

> ...brooding even here. The moose, the beaver and the bear had for years been decreasing and other fur-bearing animals were slowly but surely lessening with them. The natives, aware of this, were now alive, as well, to concurrent changes foreign to their experience. Recent events had awakened them to a sense of the value the white man was beginning to place upon their country as a great storehouse of mineral and other wealth. These events were, of course, the Government borings for petroleum, the formation of parties to prospect, with a view to developing the minerals of Great Slave Lake, but, above all, the inroad of gold-seekers by way of Edmonton. The latter was viewed with great mistrust by the Indians...

(Mair 1908: 24)

Indigenous groups saw treaties as ways that would allow them to enter into binding agreements with the British Crown and the Canadian federal government which, in turn, would allow for further settlement as well as being of benefit to indigenous people. For the federal government, however, treaty-making was very much part of a process of extinguishing Aboriginal title to lands, a way of preparing the way for peaceful settlement and agricultural development of the north-west frontier, as well as gaining access to lands with significant resource development potential. Brody (1981: 63) describes how Indians were viewed as an obstruction to settlement and development and how treaties provided a means for their removal: "If unimpeded settlement of the West was to proceed, some limitation of Indian presence was required."

The federal authorities believed that, in many parts of the North-West, hunting and trapping were the only possible activities for Aboriginal people and they ignored calls for treaty negotiations in these areas, especially if there were no opportunities for development by non-indigenous peoples. In some cases, indigenous groups were purposely excluded from treaty signings because the federal government believed it made no sense to enter into negotiations in places where there was no apparent reason, or economic motive, to do so. The Treaty 8 commission, for example, did not include the Lubicon Cree of what is now northern Alberta, and the consequences of this today will be discussed in Chapter Six. The numbered Treaties, in effect, emerged from non-indigenous ideas and interests concerning development and settlement, federal government priorities for obtaining territories and lands, and how to compensate Native people for their loss of Aboriginal title.

Treaty 8 came about primarily because of the need to open up the Peace River frontier area of north-west Alberta and north-east British Columbia to settlement. There had long been speculation about the agricultural promise of this part of northern Canada, as well as the potential for mineral development. In the 1890s, as today, prospective oil, gas and mineral reserves "were an important, if at times fanciful, aspect of dreams about the frontier" (Brody 1981: 64). As Leonard notes,

For years, the government has considered the merits of a treaty with the native peoples north of Edmonton. As early as January

1890, Indians from Lesser Slave Lake and the upper Peace River were reported to be interested in one, and, in January 1891, the Privy Council expressed its belief that the region contained sufficient mineral resources to make a treaty advisable.

(Leonard 1995: 16)

The Peace River region, as well as the other vast boreal forested areas north of Edmonton, in an area then known as the District of Athabasca, became known as "the delayed frontier" partly because the land had previously been seen as unfit for development. But by 1898 parts of the North, from the region around Edmonton and north into the Mackenzie Basin, were on the verge of large-scale development and settlement and the "richness of the soil in the Peace River region had long been publicized" (Leonard ibid.: 16). Yet Treaty 8 also became necessary because of the events in the Yukon following the discovery of gold in the Klondike in the 1890s. Edmonton, at that time a small town emerging from its origins as a Hudson's Bay Company trading post on the banks of the North Saskatchewan River, ventured to capitalize on its location at the southern edge of the Mackenzie Basin as the starting point for an all-Canadian route to the Klondike gold fields, via a transportation route using the Mackenzie River and its tributaries. Lured by the promise of gold and wealth, prospectors were heading to the Yukon via Skagway in south-east Alaska, and over the Chilkoot Pass, or were blazing other trails from southern Alaska or east along the Yukon River. As historian Pierre Berton (2001: 216) wrote, "While thousands were trying to reach the Klondike over glaciers, mountain passes, river routes, and swamps, the merchants of Edmonton were doing their utmost to convince the world that their city was the gateway to the only practicable trails."

Improvements were made in the river transportation system and soon prospectors were heading north to Edmonton en route to realizing their dreams of finding Yukon gold (the importance of the Edmonton trails was celebrated from the 1960s until the mid-2000s as a theme integral to Edmonton's history in the city's summer *Klondike Days* festival). The impact on Edmonton was profound. To quote Berton again,

Upon this backwater the gold rush burst like a cyclone. Suddenly thousands of men appeared, jamming the streets. The flats along the North Saskatchewan blossomed with tents. Sleds loaded with provisions clogged the thoroughfares. And the zaniest pieces of equipment since the days of the Ark were trundled through town. Indeed, there actually was an ark, a curious boat of galvanized iron intended for use in all seasons, with a keel for river travel and runners for snow.

(Berton 2001: 219)

Although some of the overland routes from Edmonton turned out to be "among the most impracticable" (Berton 2001: 216.), larger parts of the northern frontier were opened up and, while many failed to reach Yukon via the rivers and trails north of Edmonton, other developments following the Klondike discoveries included prospecting for minerals in the area around Great Slave Lake. Great interest was soon aroused in the potential for new extractive industries in Canada's vast North-West. Aboriginal title to lands, however, as well as a clash of cultures and social and economic disruption, threatened to interfere with prospecting and resource exploitation. Aboriginal title is historic and communal (i.e. an individual cannot hold Aboriginal title) and, as defined by the courts in Canada, it is a right to the land itself, not just the right to carry out traditional resource use practices. Mair (ibid.: 24) described the fears that, if treaties were not made in the North, indigenous peoples, "soured by lawless aggression, and sheltered by their vast forests, might easily have taken an Indian revenge and hampered, if not hindered, the safe settlement of the country for years to come". So, deciding "to treat with them at once on equitable terms, and to satisfy their congeners, the half-breeds, as well" (Mair ibid.), the federal government moved swiftly to enter into negotiations with the Native peoples in the boreal plains and foothills of the southern Mackenzie Basin, beginning with the districts north of Edmonton. In 1899, a party of some 50 government officials, guides and observers set out from Edmonton to make Treaty 8 with the peoples who had lived in these northern lands for generations. Leonard has described how, as in other parts of Canada,

the Indians were to be settled with the provision of reserves, annu-
ity payments, educational opportunities, and equipment, supplies
and training in farming. The Métis, on the other hand, were to be
settled by the issuance of scrip. This being certificates entitling
each individual to so many acres of land or many dollars towards
the purchase of land...Because the Athabasca region held so many
people of mixed blood who lived like Indians but did not know their
actual status, provision was made for them to choose either treaty
or scrip. Likewise for those Indians preferring land or money scrip
instead of reserves and their attending benefits, allowance was
made for them to take such scrip instead of treaty.
(Leonard ibid.: 16-17)

Sharing the Land and the Sacred Nature of Treaties

The discussions and negotiations surrounding the treaties in the
Canadian North-West are complex and some aspects remain contentious but, nonetheless, detailed analysis is beyond the scope of
this chapter. For the Peace River and Athabasca districts, Leonard
provides a good summary of the process of the negotiating and
signing of Treaty 8 as well as the uncertainty and concerns expressed by the Indians and Métis, and Charles Mair's *Through the
Mackenzie Basin* is an excellent first-hand account of its signing and
the distribution of scrip. Of particular interest, given the importance of Aboriginal voices in environmental impact assessments
and public hearings processes today (from the Berger Inquiry on-
wards), is Mair's recording of the words of the Commissioners and
the Aboriginal Chiefs. One, Keenooshayo, the chief spokesman for
the Cree, expressed his feelings of how unsure he was about the
purpose and nature of the Treaty:

Do you not allow the Indians to make their own conditions, so that
they may benefit as much as possible? Why I say this is that we
to-day make arrangements that are to last as long as the sun shines
and the water runs. Up to the present I have earned my own living
and worked in my own way for the Queen. It is good. The Indian

loves his way of living and his free life. When I understand you
thoroughly I will know better what I shall do. Up to the present I
have never seen the time when I could not work for the Queen, and
also make my own living. I will consider carefully what you have
said.

(Keenooshayo, quoted in Mair ibid.: 60)

Originally published in 1908, Mair's book reflects his ardent Canadian patriotism, belief in British institutions, and his proud support of national ideologies of development and breaking down the frontiers of the north-west of Canada. For Mair, the North was just as eastern Canada had been 30 years before, a land of great resource potential waiting to be opened up and developed: "There is fruitful land there," he wrote, "and a bracing climate fit for industrial man, and therefore its settlement is certain" (Mair ibid.: 148). Yet despite its overall tone supporting Canadian imperial advancement into the remote North-West, *Through the Mackenzie Basin* remains a powerful documentary source for understanding a significant piece of the history of relations between Aboriginal peoples and the government in Canada, and for understanding the contemporary relevance of treaties.

For First Nations in Canada struggling to seek a voice in megaproject development, the treaties made with the Crown have a sacred nature, as well as being political and legal documents. Venne argues that it must be remembered that the Crown came to indigenous peoples to make treaties, and that indigenous notions of land and traditional legal systems guided the ways in which the Chief negotiated and concluded the treaties:

The Cree did not go to England to make treaty. The Cree Peoples did not go to Ottawa. The Crown sent its representatives to our lands. There was no conquest on Cree territory. There was no war with non-indigenous people. Our territories were not terra nullius ("land of no-one"), because we were here. As Nations, we had our own governments, our own laws, our own political and legal systems operating in our territories.

(Venne ibid.: 3)

Venne urges us to remember that both indigenous and non-indigenous peoples have treaty rights. The fundamental aspect of Treaties 6, 8 and 11 is that indigenous peoples agreed to share the land with non-indigenous people, so the treaty rights for the latter are that they can live on Aboriginal lands in north-west Canada. As indigenous leaders point out time and time again, indigenous peoples continue to honour that right and do not interfere with the treaty rights of non-indigenous peoples (IWGIA 1997, Venne ibid.). At the same time, they often argue that non-indigenous peoples have reneged on treaty rights.

Although writing about Treaty 6, Venne (1997) points out that the Chiefs and Headmen only agreed to share the top soil to the depth of the plough and non-indigenous people, in return, agreed to provide certain benefits. In addition, areas of land were reserved for the exclusive use of indigenous people and they were also allowed certain hunting, trapping and fishing rights on Crown lands. There was no surrender or selling of land by indigenous peoples when they signed the numbered treaties of the North-West, according to many contemporary First Nations, despite the fact that the Indians did, as the words of Treaty 8 emphasize, "hereby cede, release, surrender and yield up to the Government of the Dominion of Canada, for her Majesty the Queen and her successors for ever, all their rights, titles and privileges whatsoever to the lands included in the following limits…".

Understanding the continuing importance of treaties, and how First Nations view those treaties, allows us insight into the reasons why struggles to regain control over lands and resources are often a source of conflict and tension between Canada's Aboriginal peoples and federal and provincial governments. Brody (ibid.) describes how, in testimony to the Alaska Highway Pipeline hearings in 1978, legal experts argued that the written records of the negotiations revealed that the Indians did not understand Treaty 8 to be a surrender or transfer of rights but that it was a treaty of peace and friendship. The theme of Treaty-signing with the Crown was also an important element of stories told at the Berger Inquiry (Scott ibid.). Calliou (ibid.) argues that there remain divergent and competing interpretations of what these historical treaties actually mean, with Crown officials viewing Treaty 8 (and other treaties) as an extinguishment of Aboriginal title and a transfer of land, while First Nations leaders themselves continue to insist that treaties remain important as a "nation to nation" agreement to share their lands and their resources, not an extinguishment of title or rights. Yet for the government and the agents of the Crown, claims Brody, the treaty was a way of restraining the indigenous population, while giving them certain reassurances. Above all, it was a way of protecting "the white man's frontier (whenever or wherever it might need to be) against possible limitation in the future" (Brody ibid.: 64).

Gray (1997: 20) has argued that, "The parties to the Treaties had different interests which were brought together in the documents. The conditions under which the documents were drawn up, the various interpretations of the clauses and their significance now make them as relevant today as when they were signed." The agents of the Crown saw things differently from Aboriginal peoples. They thought they were acquiring land for the Crown and ownership over resources. Two important Canadian Supreme Court decisions emphasize Aboriginal rights and title. The first, known as the Calder Decision of 1973, ruled that Aboriginal people in Canada have an ownership interest in their ancestral lands and resources, and it also pointed out that rights could only be extinguished if Aboriginal people had knowingly surrendered them. It was a pivotal decision and led to the Canadian government introducing a land claims policy. In the Delgamuukw Decision of 1997, the Supreme Court held that Aboriginal title was an Aboriginal right that is recognized and affirmed in Section 35(1) of the *Constitution Act, 1982*. Its significance lies in its acknowledgement that Aboriginal title provides First Nations with a right to land and that they have entitlements which allow them to practise resource use activities on traditional territories. Furthermore, if First Nations have a valid claim under Canada's Comprehensive Land Claims policy, they have the right to say how the land is used until that right is extinguished knowingly. As I will discuss in Chapter Four, with particular reference to the Dene Tha' and Dehcho First Nations, the importance of the continuing political and legal power and the cultural relevance of treaty rights in northern British Columbia, northern Alberta and in the central Mackenzie Valley of the Northwest Territories, as well as the sacred nature of treaties themselves, are often evoked by Aboriginal peoples in discussions of oil and gas development, particularly proposals for pipeline construction, and during environmental hearings for hydro-electric development.

The Duty to Consult and the Right to Benefit

In 1998, the Canadian government transferred responsibility for oil and gas in Yukon Territory to the territorial government. Following land claim settlements, Sahtu, Gwich'in and Inuvialuit beneficiaries in the Northwest Territories, and Inuit beneficiaries in Nunavut, became holders of private mineral and surface rights over defined blocks of land within their respective settlement regions. As a result, they now manage their own petroleum rights on these lands. As discussed earlier, the management and development of oil and gas resources on Canada's federal lands in the Northwest Territories, Nunavut and Arctic offshore areas is a federal responsibility, overseen by the Northern Oil and Gas Directorate of Indian and Northern Affairs Canada.

If proponents of energy projects are to be successful in obtaining regulatory approval to explore and develop oil and gas, and to construct and operate pipelines, they must have a good understanding of northern Canada's complex social, cultural and political environment and be able to navigate and negotiate it successfully. Seismic trails and pipelines must pass through, and oil and gas wells are often drilled on, Native lands which may be subject to historic treaty, a modern land claims settlement or an outstanding land claim. To consult and deal with Aboriginal communities may be a statutory requirement, but it is also a practical business matter. In the Northwest Territories, comprehensive land claim agreements require that project proponents enter into certain forms of agreement with the beneficiaries over specific issues. Different types of agreement may be entered into depending on the type of land claim agreement or treaty that applies in the area where the planned development is to take place, be it seismic data gathering, pipeline construction, well drilling, and so on.

Before initial applications to explore for oil and gas on land in the Northwest Territories, Nunavut and as well as in northern offshore areas are submitted to regulators, energy companies have a duty to consult with communities about the proposed project and to identify areas of environmental sensitivity, such as important hunting, trapping and fishing areas, critical wildlife habitat, and sites of cultural and spiritual significance. The project proponent must also ensure that Northerners and northern businesses have full and fair access to employment, training and business opportunities. Ideally, although it does not always happen in practice, qualified northern residents and northern businesses must also be given first consideration. As the body responsible for overseeing the development of oil and gas reserves on federal lands in the NWT, INAC's Northern Oil and Gas Directorate has also stated that it will not open lands for bidding in areas where Aboriginal people do not want it and have made it clear that if an oil and gas sector is to become an integral part of the northern economy, the industry and government partners must work with Aboriginal people to strengthen northern communities.

Aboriginal households and communities in northern Canada are characterized by a blend of formal economies (e.g., involvement in commercial harvesting of fish and other animals, oil and mineral extraction, and tourism) and informal economies (e.g., harvesting renewable resources from land and sea primarily for household consumption). The ability to carry out harvesting activities is not just dependent on the presence of animals in traditional hunting areas but on the steady availability of cash, as the technologies of modern harvesting activities are extremely expensive in remote and distant northern communities. In the mixed economies that characterize northern Canadian Aboriginal communities, a half or more of household income may come from wage employment, simple commodity production, or from government transfer payments (Nuttall et al. ibid). People move between different spheres, between subsistence and market and employment activities, depending on opportunities and preference (Usher et al.

2003). Such increasing reliance on other economic activities does not mean that production of food for the household by traditional means has declined in importance or disappeared. Hunting, trapping, gathering and fishing are activities mainly aimed at satisfying the important social, cultural and nutritional needs, as well as the economic needs, of families, households and communities. Usher et al. (ibid.: 177) argue that the household is the basic unit of production and consumption in Aboriginal communities. In this social context, sharing the products of hunting and fishing with elders and community members is paramount. Sharing is important to the local economy in terms of networks of distribution, but it is a fundamental part of traditional activities that relate to identity, culture and community. Research points to the continued importance of harvesting activities despite a growing proportion of the population of indigenous communities not being directly involved in harvesting (Nuttall et al. ibid.).

While food procured from renewable resource harvesting continues to provide northern peoples with important nutritional, socio-economic and cultural benefits, finding ways to earn money is a major concern in many northern communities, where employment opportunities are limited. The interdependence between formal and informal economic sectors, as well as the seasonal and irregular nature of wage-generating activities (such as tourism) means that families and households are often faced with a major problem in ensuring a regular cash flow. Recent research also points to the reality that northern communities face serious issues in maintaining food security (Ford and Berrang-Ford 2009).

It is within this socio-economic context that oil and gas development is viewed as an opportunity for providing the potential for employment and prosperity for northern communities. The seasonal nature of employment in the oil and gas industry, particularly during winter exploration activities, is seen to fit well with the mixed traditional economy/wage economy of many communities. There is no doubt that exploration and development in the Mackenzie Valley have created some employment, training and business opportunities for Northerners and northern firms. While the communities of Colville Lake, Fort Good Hope and Tulita have seen significant oil and gas exploration activity in recent years, this has been accompanied by increased levels of local employment and business contracting. In addition to the economic benefits, the positive environmental effects of northern energy development are often cited by those in favour of opening up Canada's lands and waters to development.

Yet, as I have pointed out earlier in this book, these potential benefits to northern communities and regions do not ease local anxieties over the long-term effects of oil and gas development. As the world looks increasingly to the North for oil and gas to meet its ever increasing energy demands, the nature of these anxieties are being expressed in public hearings. This will be explored in the next chapter, which discusses public participation and Aboriginal voices in environmental assessments and regulatory reviews through an examination of the Mackenzie Gas Project.

CHAPTER IV

THE MACKENZIE GAS PROJECT AND CANADA'S ENERGY FUTURE

In October 2004, energy companies submitted applications for construction and operating permits for a Mackenzie Valley pipeline route and other associated facilities in the Northwest Territories as essential elements of the Mackenzie Gas Project (MGP). The application was filed by Imperial Oil on behalf of five partners— in addition to Imperial, the project proponents are ConocoPhillips (North) Limited, Shell Canada Limited, ExxonMobil Canada Properties, and the Aboriginal Pipeline Group (APG). On 23 November 2005, Imperial informed the National Energy Board (NEB) that the proponents of the MGP were prepared to proceed to public hearings on the applications and, following pre-hearing planning conferences held in Inuvik, Yellowknife, Fort Good Hope and Fort Simpson in the NWT in December, public hearings for the MGP were carried out throughout 2006 and 2007 as part of the regulatory review process. This process was the responsibility of the National Energy Board (NEB), which focused on the economic, technical and engineering aspects of the project, and the seven-member Joint Review Panel (JRP), which considered the environmental aspects.

All stakeholders with an interest in seeing it go ahead concur in that the project has tremendous potential benefit for the Canadian North, as well as for the economy of Alberta and other parts of Canada, and more than 30 Aboriginal groups have signed a Memorandum of Understanding with the private sector under the umbrella of the Aboriginal Pipeline Group (APG). Prior to the public hearings, the NWT government argued that the pipeline represented employment, investment and business development and the opportunity for residents of the NWT to provide a good standard of living and quality of life for themselves and their families (Jaremko 2005a). The structure for the review of the Mackenzie Gas Project, in particular how northern operators and regulators would co-operate with national regulators, was first articulated in *The Cooperation Plan for the Environmental Impact Assessment and Regulatory Review of a Northern Gas Pipeline Project through the Northwest Territories* (otherwise known as the "Cooperation Plan") a document released in June 2002. This plan involved input from the various government departments and agencies and northern regional boards responsible for the assessment and regulation of energy development in the NWT, including Indian and Northern Affairs Canada (INAC), northern boards such as the Mackenzie Valley Land and Water Board, the Inuvialuit Land Administration, the Mackenzie Valley Environmental Impact Review Board (MVEIRB), and many other boards which have been created as a result of the finalization of land claims agreements between the Government of Canada and Aboriginal groups in the Mackenzie Valley and in the Mackenzie Delta-Beaufort Sea region. Such boards include the Environmental Impact Review Board for the Inuvialuit Settlement Region, the Inuvialuit Land Administration, the Inuvialuit Game Council, the Sahtu Land and Water Board, and the Gwich'in Land and Water Board. The process reflects the institutional arrangements for environmental assessment, consideration of development projects and environmental management that have evolved in the Northwest Territories over the last thirty years.

At the opening of the NEB hearings, Fred Carmichael, President of the Gwich'in Tribal Council and Chairman of the Aboriginal Pipeline Group, expressed a common sentiment that Aboriginal peoples in the NWT were ready for oil and gas development. The social, cultural, economic and political context in this part of northern Canada is now very different from the time when the proposal for a pipeline was initially reviewed by Thomas Berger. Land claims and increased self-governance, as well as the need for economic development and the creation of jobs for Northerners, all means that there is a more positive perspective on industrial development generally found in some parts of the NWT. In addition, indigenous communities have had experience of working with, as well as deriving some benefits from, oil and gas companies, as well as with the diamond mining industry. However, despite this changed mood, not all in Canada's North (to say nothing of opposition from southern-based environmental groups) are in agreement with the current plans for gas development in the Mackenzie Delta and the construction of a pipeline along the Mackenzie Valley. Land claims and negotiation issues between the federal government and the Dene communities represented by the Dehcho First Nations in the central Mackenzie Valley are ongoing and remain unresolved; a lawsuit was filed in 2006 by the Dene Tha' people of northern Alberta, who felt excluded from the consultation and regulatory review processes and were concerned with the infringement of Aboriginal and Treaty rights; and there are many concerns about environmental and social impacts voiced by environmental NGOs and by various Aboriginal groups and individuals.

The settlement of comprehensive land claims and the greater recognition of indigenous rights in Canada's North has meant that benefits from resource development projects on Aboriginal-owned lands should be secured through participation agreements, as well as resource revenue sharing and equitable access to government contracting and economic programmes. Major development projects planned by the oil and gas industries nonetheless raise the prospect of far-reaching social, economic and environmental changes for Aboriginal peoples and northern ecosystems. In this chapter, I focus on the Mackenzie Gas Project as a case study which highlights this, as well as some of the concerns and views of Aboriginal peoples in the Northwest Territories and northern Alberta. I look at some of the most prominent issues arising from discussions surrounding this controversial project (which some government and energy company officials consider will decide the near future of energy development in Canada's North), examine local concerns over participation and consultation, and show how it provides insight into some of the contested perspectives on the future of northern Canada, its peoples and the environment.

A commonplace remark expressed by many people in the Northwest Territories is, "This pipeline will change the North forever." Some speak of this in a positive way, talking of their hopes for the future of their communities and regions, others are concerned about irreversible impacts and have spoken out against the pipeline and its associated developments; some see the Mackenzie Gas Project as an unprecedented opportunity to plan for a sustainable northern economy in the NWT and to control and manage the cumulative impacts of resource development; many others are simply resigned to the inevitability of development, whatever their opinion may be (for example, speaking at a JRP hearing in Inuvik on 17 February 2006, Ruby Koe said, "Whether you like it or not, it's going to come. We can't do nothing about it."). These opinions, concerns and aspirations have been expressed extensively at both the NEB and JRP public hearings and have been recorded along with other submissions and testimony in thousands of pages of transcripts (there are over 11,000 pages of JRP transcripts alone) which are available in a public registry on the website of the Northern Gas Project Secretariat (http://www.ngps.nt.ca).

The Mackenzie Gas Project

The Mackenzie Delta is, after Russia's Lena River Delta, the second largest Arctic delta, a vast system of lakes, ponds, meandering rivers, channels, tidal mudflats, peat bogs and low-lying islands. It is dotted with countless tundra polygons and around 25,000 lakes cover 25% of its total surface area. At some 1,800 km in length, the Mackenzie River (called *Deh cho*, or "great river" by the Dene) is the main branch of the second largest river system in North America (after the Mississippi-Missouri river system). The watershed of the Mackenzie River (named after Sir Alexander Mackenzie, who became the first to descend the river to the Arctic Ocean in 1789) is called the Mackenzie Basin. A northern extension of the North American Great Plains, flanked by the Rocky Mountains to the west and the Canadian Shield in the east, the Mackenzie Basin includes several major rivers (including the Peace, Athabasca and Slave) and three major lakes (Lake Athabasca, Great Slave Lake and Great Bear Lake) and drains approximately 20% of Canada. Like other parts of the Canadian western Arctic, the Mackenzie Delta is a land relatively rich in resources. In some ways, it is an oasis bordering on High Arctic deserts. This is critical habitat for wildlife and an ancestral homeland for indigenous peoples. The Mackenzie Basin is the traditional territory of Inuvialuit, Gwich'in, Dene and Métis indigenous peoples in the NWT, and Dene and Cree in northern Alberta.

As in many other parts of the North, fur trading was the primary economic activity in the Mackenzie Basin from the early 19th century until the 1930s. Dramatic changes also came to the Mackenzie Delta and other Inuvialuit communities in the Beaufort Sea region when American commercial whalers arrived in the second half of the 19th century and based many of their activities on Herschel Island. While the fur trade shaped the livelihoods of indigenous peoples and contributed to the fortunes of outsiders who came to the region, it was never going to be a sustainable activity—a combination of overexploitation of fur-bearing animals and changing global fashion tastes saw to that. It is an example of the boom and bust nature of non-renewable resource development often experienced by northern indigenous peoples and circumpolar environments since their incorporation into the global economy.

The Mackenzie is one of the last great free-flowing rivers anywhere in the Arctic, although there are plans for hydro-electric development that will challenge its ecological integrity. And while tourist brochures describe the Northwest Territories as one of the last great wilderness areas left in North America, evidence of human disturbance persists in the environmental footprint of seismic trials left behind by geological surveys for oil and gas in the 1960s and 1970s. The full extent of previous exploration activity is best viewed from the air on a flight over the Northwest Territories, and the Mackenzie Valley in particular. Scars in the form of clear-cuts and cut-lines, which were created by removing trees, shrubs and other vegetation, in addition to the well-sites left by seismic exploration and drilling crews, can be seen crossing the land in a grid pattern at intervals of several hundred metres. Extending in some cases close to the Arctic coast, these seismic lines are often five to eight metres across and appear from the air as bare, straight narrow strips stretching into the distance.

As is true for many other Arctic regions, the popular image of the Mackenzie Delta is of an untouched, pristine wilderness. Yet its history tells us a compellingly general story about indigenous peoples, the changes they have experienced and endured in recent times, how outsiders have viewed the Arctic, and how both indigenous and non-indigenous peoples imagine its future. The untapped natural gas reserves of the Mackenzie Delta are estimated at some 55 trillion cubic feet. Natural gas is already produced for local consumption and is piped via the 60 km-long Ikhil pipeline to Inuvik, a town of some 3,300 (mainly Inuvialuit, Gwich'in and non-Aboriginal people) and the administrative centre for the Mackenzie Delta and the Western Arctic. Inuvik was a boom town in the 1970s. Imperial and Shell established base camps and Eddie Kolausok, an Inuvialuit land claims negotiator, describes how mechanics, electricians, welders and a host of other tradespeople moved into Inuvik. Winter roads and airstrips were built to transport the seismic teams and the crews of drilling rigs. "Inuvialuit got jobs as equipment operators, cooks, camp attendants, roughnecks, derrick hands, bear monitors, expediters and truck drivers. People who were used to driving dog teams could now afford snowmobiles." But this boom ushered in an inevitable process of social impacts:

> *It brought work, money and many southern transients. Inuvik's bars were often rocked with scenes of drunken conflict. Young people dropped out of school to take high-paying—but temporary— jobs in the oil industry. Violent assault, break and enter, theft and suicide all increased. Drugs and sexually transmitted diseases appeared. Even trappers living far from town would sometime come face to face with oil workers moving heavy equipment across their traplines.*
>
> (Kolausok 2003: 177)

Following the Berger Inquiry, the oil companies began to pull out of Inuvik in the late 1970s and the oil boom came to an abrupt halt. But things were not quiet for too long. Exploration and drilling camps have been increasingly active in the Mackenzie Delta since the late 1990s, as energy companies gauge the potential for future development. Preliminary estimates suggest employment for as many as 2,600 short-term positions during the construction phase of the Mackenzie Gas Project, as well as 50 permanent, long-term jobs related to the Mackenzie Valley pipeline and other facilities during the operational phase. The anchor field development promises yet more employment, with construction, drilling and servicing and operations staff required for the project. NWT government officials, optimistic that the territory will reap the benefits from non-renewable resource development in the same way as Alberta has done from its oil and gas industry, point to the economic growth beyond the immediate job offerings. As one territorial government official stated, "This is an undeveloped part of Canada, and the benefits will not just come from the gas industry but the NWT will finally get some infrastructure that will allow other industries and support services to operate here. The Mackenzie Gas Project will really open up the North."

Government officials have also gone on record as saying that the MGP represents "the last chance" for Arctic gas (Jaremko ibid.). In the Northwest Territories, the period between 2005 and 2007 was dominated by discussion over the regulatory process, the procedures for the technical, environmental and social assessment for the Mackenzie Gas Project, as well as the nature of the public hearings process. Concerns were also expressed by industry that demands for compensation and benefits from communities were far higher in the NWT than in industry operations elsewhere, such as Alberta, and prolonged attempts to make land access and benefits agreements had the effect of creating negative northern views of the project. In particular, critics of the project have argued that the corporate partners were not interested in proposing alternative investments if the MGP did not happen (Jaremko 2005b). The Inuvialuit have demanded that socio-economic issues such as education and housing should be settled before the pipeline is built, while the Sahtu Dene argue that the energy companies must realise that, in negotiating with Aboriginal people, they are dealing with governments not landowners. The public hearings, however, only began following a period of around three years during which the proponents engaged in public consultation, carried out traditional knowledge studies, conducted technical engineering and environmental studies, assessed the impacts on local communities and de- ~

veloped northern benefits plans that address education, training, employment and business opportunities.

The Mackenzie Gas Project comprises several elements. A gathering pipeline system will connect three natural gas production anchor fields in the Mackenzie Delta—Taglu (Imperial), Parsons Lake (ConocoPhillips, ExxonMobil) and Niglintgak (Shell)—to a gas processing facility near Inuvik, where the gas and liquids will be separated. From there, gas will be transported by a 30-inch 500 km natural gas liquids pipeline to Norman Wells on the Mackenzie River. Continuing from Norman Wells, a 30-inch buried dry gas transmission pipeline of 800 km will parallel an existing oil pipeline to northern Alberta and will connect to the natural gas pipeline system operated by TransCanada Pipelines. Compressor stations will also be built at intervals along the route. The proposed project crosses four Aboriginal regions in the Northwest Territories (the Inuvialuit Settlement Region, the Gwich'in Settlement Area, the Sahtu Settlement Area and the Dehcho Territory). A short segment will be in north-western Alberta near the NWT border. It is a multi-year phased project and stakeholders had originally hoped for gas production to start between 2008 and 2010. However, with a series of delays associated with the hearings and negotiations between federal and territorial governments and Aboriginal groups, if the MGP is approved the pipeline may not be in service until 2013 or even later. The three anchor fields supplying the gas can generate about 800 million cubic feet per day. The pipeline will be designed for 1.2 billion cubic feet per day as the proponents hope that future development in the Mackenzie Delta and the Colville Hills area will add more gas to the pipeline. The total length of the natural gas pipeline will be about 1,300 kilometres and it is this pipeline that is at the centre of controversy and debate, so much so that the other elements of the Mackenzie Gas Project are often forgotten.

The regulatory hearings process comprised a series of hearings about the nature of the technical and engineering aspects of the project and these were conducted by the National Energy Board (NEB) along with parallel hearings on environmental, social and economic issues, which were conducted by the Joint Review Panel

(JRP). The JRP comprises seven members, including four Northerners who are resident in the North. The JRP has its origins in 2004, when the Canadian federal Minister of the Environment, the Mackenzie Valley Environmental Impact Review Board (MVEIRB) and the Inuvialuit Game Council concluded an *Agreement for an Environmental Impact Review of the Mackenzie Gas Project*, otherwise known as the Joint Review Panel—or JRP— Agreement. As well as specifying the mandate of the JRP and the scope of the environmental impact assessment, it established the process by which the JRP members were chosen. Four members were chosen by the Inuvialuit Game Council, three members by the MVEIRB, and one by the Minister of the Environment. The JRP's task has been to review those documents that have been submitted by the project proponents, examine the environmental impact statement and incorporate all of the input received from participants at the public hearings. The JRP was mandated by the environmental assessment authorities to place considerable importance on traditional knowledge relating to the environment, the land, the animals and fish that inhabit the land and waters of the proposed pipeline routing. For the purposes of the environmental assessment, the JRP's task was to consider the environmental impact of the MGP as well as the connecting facilities in northern Alberta. As I will discuss later in this chapter, the review of the latter has caused considerable confusion and controversy. It is also important to note that, while the JRP's mandate is to consider environmental effects of the MGP, including negative and adverse impacts on Aboriginal activities and livelihoods, the JRP was not given a mandate to conduct Aboriginal consultation and cannot consider the legal aspects of Aboriginal rights or land claims.

Ken Vollman of the National Energy Board opened the hearings in Inuvik on 25 January 2006 by saying that the members of the NEB panel

are pleased to be in Inuvik today to begin hearing directly from those who will be most affected by what I think is fair to call a historic undertaking. We're striving to hear all voices and also make participation by Northerners as easy as we can.[19]

Throughout 2006, the hearings were then carried out in 26 communities in the Northwest Territories, along with communities in Alberta. Originally, it was planned that the hearings would conclude in December 2006, with the Joint Review Panel aiming to submit its report sometime during the spring of 2007. However, the JRP extended its hearings until November 2007, as it determined that considerably more testimony and evidence needed to be heard and gathered. The Joint Review Panel finalized and submitted its report to the NEB in December 2009. The JRP has endorsed the project, saying that it "offers a unique opportunity to build a sustainable future in the Mackenzie Valley and Beaufort Delta regions. The Project itself, as long-term infrastructure, provides a key basis for future economic development. This opportunity carries the risk of adverse impacts, however" (Joint Review Panel for the Mackenzie Gas Project 2010: v). The JRP report lists 176 recommendations, all of which it argues should be fully implemented if the MGP is to deliver valuable and lasting overall economic benefits while avoiding significant adverse environmental impacts. It acknowledges that the project would be the occasion for major change throughout the region. Some of the JRP's recommendations include a range of measures to enhance socio-economic benefits, such as training programmes, reducing barriers to employment that relate to gender and diversity equality, minimizing the impacts of in-migration, and dealing with the impacts of alcohol and drug abuse. Other recommendations concern environmental protection, but the JRP has recommended that the Government of Canada commits "the funding required to implement things it has already committed to do, such as fulfilling its obligations under the *Species at Risk Act*, the *Mackenzie Valley Resource Management Act*, and the Protected Areas Strategy" (Joint Review Panel for the Mackenzie Gas Project ibid.: vi). Given the long list of recommendations, some argue that the JRP report provides an outline for a plan for responsible and sustainable development in the Northwest Territories (e.g. Grant 2010), although the National Energy Board proposed modifications to some of the JRP's recommendations.

The National Energy Board has had the enormous task of reviewing the testimony and all information presented by the proponents, interveners and communities at both sets of hearings, as well as the JRP report. Following the release of the JRP report, all parties to both sets of hearings were invited to respond to the panel's recommendations and, almost four months later, in April 2010, the NEB held a final round of hearings (called a final argument) in Yellowknife and Inuvik to receive final input, opinions, concerns and arguments about all the evidence that had been submitted, including the JRP report. The NEB is expected to release its decision to the Canadian government sometime in late 2010. If approval is given to the proponents to construct the MGP, the Mackenzie Valley Land and Water Board and the Northwest Territories Water Board will hold their own public hearings, although this process will not be as extensive or as long as the federal review. If final approval is eventually given for the Mackenzie Gas Project to proceed, the regulators will issue the necessary permits and licenses. The most important of these is a "Certificate of Public Convenience and Necessity" (CPCN), which marks the completion of the regulatory process and is the main final permit required by the proponents to move forward with the project. The federal Cabinet must approve the actual issuance of a CPCN.

Aboriginal Participation: the Aboriginal Pipeline Group

Aboriginal peoples are major stakeholders in the Mackenzie Gas Project. With most Aboriginal groups in the NWT having had their land claims settled in the 1980s and 1990s, a milestone meeting took place in Fort Liard in the NWT in January 2000. Aboriginal groups met to discuss the prospect of oil and gas development for the first time since the Berger Inquiry. Rather than telling industry not to build a pipeline on Aboriginal lands, the participants discussed how they might be involved in a pipeline project. As Fred Carmichael told the Joint Review Panel in Inuvik in February 2006, "At that time, the decision was made that if there were going to be a pipeline through our territory, we would want some ownership in order to maximize the benefits to our people."[20] Following on from the meeting, the leaders of the Inuvialuit, the Gwich'in and the Sahtu Dene formed the Aboriginal Pipeline Group (APG) and partnered with Imperial Oil, ConocoPhillips, Shell Canada and ExxonMobil in the Mackenzie Gas Project consortium.

Essentially a business venture owned and controlled by NWT Aboriginal groups, the idea behind the APG is to offer a new model for Aboriginal participation in the developing economy of the NWT, to maximize Aboriginal ownership of development projects and benefits from the proposed Mackenzie Valley pipeline, and to support greater independence and self-reliance among Aboriginal people. It has been praised by governments and industry as an example of how Aboriginal people in Canada are making a contribution to the country's economic development and competitiveness. If the pipeline is built, the APG will eventually be one-third shareholders in it. Imperial Oil has a 34.4% ownership. Conoco Phillips has 15.7%. Shell Canada has 11.4%, while Exxon Mobile Canada has 5.2%. During initial negotiations, the APG attempted to obtain the highest percentage of project ownership, but the economics of participation meant that higher ownership based on the APG's financial model was not possible. The business deal that was eventually concluded aims to work as follows: the Aboriginal Pipeline Group would plan to finance 100% of its investment with no risk to Aboriginal peoples, the APG would obtains loans to finance that investment, producers would sign long-term shipping contracts, loans would be re-paid from the APG's share of the pipeline revenue, and the balance of the APG's revenue would be returned to APG shareholders as dividends.

Initially, however, and if the project is approved, during the construction and operating phases the APG ownership will be directly proportional to the amount of gas that will be eventually shipped through the pipeline. With new discoveries anticipated in the Mackenzie Delta, the APG would expect to receive additional capacity, which will eventually increase its ownership percentage to 33.33%. Therefore, the APG share is adjusted to reflect actual shipping commitments on the pipeline and the group is expected to increase this amount ten years after gas production has begun. Because of this, the future of the APG is linked to the success of exploratory companies.

The way benefits would be distributed within the different Aboriginal regions is outlined in each group's land claims agreement. The tax revenue rights go to whichever group has jurisdiction over that land. On Aboriginal land where they have surface and subsurface rights, Aboriginal people would receive the taxes, whereas on Crown land tax revenues would go to the Crown. But this situation also depends on devolution and ongoing negotiations with different communities. Questions remain, however, as to how to offset the lack of revenue sharing over the next few years, and how to distribute this between the Government of the Northwest Territories and Aboriginal groups.

The Aboriginal Pipeline Group sees participation in the Mackenzie Gas Project as a "win-win" situation, with a possible greater share in the future. Aboriginal attitudes have thus changed significantly since the Berger Inquiry, exemplified by the activities and perspectives of key Aboriginal leaders. Nellie Cournoyea, leader of the Inuvialuit land claim organization in the Mackenzie Delta and a former NWT Premier, lobbied against pipeline development in the 1970s. Now she is one of its most vocal supporters. One big difference between now and then, as leaders like Cournoyea point out, is that there was previously no real desire on the part of industry or government to think of Aboriginal people as meaningful participants in the pipeline project. Stephen Kakfwi, former premier of the NWT and now a negotiator for the Sahtu Dene said of the 1970s Mackenzie Valley pipeline proposal:

As a young man I worked with the Dene Nation to block the development of this pipeline. The Aboriginal people of the Northwest Territories were opposed to this project because we recognized that this development would not benefit our people. A generation ago, the Aboriginal people of the Northwest Territories were not involved in the development of the pipeline proposal. We were not consulted. We were not included in the decision-making. We were also just embarking on the vitally important process of negotiating Aboriginal land and resource rights throughout the Mackenzie Valley.[21]

One other crucial difference is that Berger wrote his report at a time when Aboriginal communities in the Mackenzie Basin were still largely dependent on trapping, hunting and fishing. At the start of the Joint Review Panel (JRP) hearings in February 2006, Fred Carmichael laid out his arguments as to why the pipeline had to be built: "A pipeline down the Mackenzie Valley will…not destroy the land, but without some form of economic base we will surely destroy our people." Carmichael said, "Some people wonder why I support this project," and went on to explain by giving a short history of the changes he had witnessed over the past 60 years:

I was born in Aklavik, a small community about 40 miles to the southwest across the Delta and raised on a trapline. In my late teens, I left trapping to become a commercial pilot. And as a bush pilot, I flew across these lands for 50 years. And I have witnessed the changes, not only changes to our way of life, but also the changes in the way industry treats our land. In the '60s and '70s, the exploration companies seemed to have little or no respect for our lands, and some of the scars are visible today. An example is the seismic lines you see crisscrossing the country. The land and environment is very important to our people. However, over the years through land claims and the resulting regulatory boards, and agencies and new technology, there have been great improvements in how we protect the land and environment. Today there is an understanding and a respect between industry and Aboriginal people. Furthermore, we are better educated and equipped to deal with industry on a level playing field, having approximately 30 years, since Berger, to prepare ourselves. And I think the fact that the Aboriginal people are a partner in this project and the fact that I'm at this table representing Aboriginal people tells you how far we've come in that 30 years. So, my friends, I want to tell you that like many of our people, I came from a trapping economy to a cash economy. Just 40 short years ago, Aboriginal people had their own economic base, which was the trapping industry. We're independent, proud and self-sufficient. This trapping economy was destroyed by people or organizations that either did not understand or care that this was our livelihood they were killing. As a result, we were forced to depend on a cash economy over a very short period of time. As a result, many of our people have had to become dependent on government and the social welfare system. Today our people are looking for a way to become self-sufficient again. We realize for this to happen we must have an economic base. As there are no other industries in this area, such as mines and so on, we see this opportunity in oil and gas and pipeline development as a way to provide that economic base.[22]

Stephen Kakfwi has also spoken of the pipeline as being a way for the Sahtu to extricate themselves from "Third World conditions" (Ebner 2005). At the JRP hearings in Fort McPherson on 17 February, Chief Charlie Furlong spoke of his hopes for economic independence in the wake of the pipeline:

The royalties, the taxes that will be generated from exploration and the pipeline will give us that independence. If we are to rebuild as a nation, then we must take advantage of economic opportunities to build our own source revenue that will allow us to be truly self-governing and perhaps one day be the proud nations our grandparents talked about.[23]

The Berger Inquiry expressed concern that Canadian Aboriginal people would not benefit economically from the Mackenzie Valley pipeline and he emphasized the importance of recognizing that they should have more control over development in the North. At that time, the NWT was more strongly divided in terms of outside business interests *versus* Aboriginal interests than it is today. Comprehensive land claims agreements have made Aboriginal business ventures in the Northwest Territories closer to their counterparts in Alaska than elsewhere in northern Canada. When the U.S. government settled land claims with indigenous people in Alaska in 1971, 13 regional Alaska Native corporations were established under the Alaska Native Claims Settlement Act (ANCSA). Today, many of these corporations are involved in some way in the oil and gas industry. On the North Slope, the Ahtna Construction and Primary Company is involved in oil spill response and pipeline work.

The Arctic Slope Regional Corporation includes Alaska Native-run oil and gas companies, and Doyon Ltd. and Cook Inlet Region Inc. both provide oilfield support services. In terms of training, the First Alaskans Institute offers summer internship positions that can be in the oil and gas sector. Northern Canadian Aboriginal-owned corporations resemble Alaskan Native corporations in both their institutional culture and business-ambitions. Since the 1980s, opportunities for Aboriginal participation in the oil and gas industries in the North have increased significantly. In the Mackenzie Delta, a variety of Aboriginal-owned companies operate from Inuvik. Such businesses include the Inuvialuit Development Corporation, which has one-third ownership in the Ikhil project, wells, processing facilities and pipelines through a Can$30 million joint venture with AltaGas Services Inc. and Enbridge Inc., and the Inuvialuit Petroleum Corporation which has been successful in developing oil and gas in southern Canada and is a major player in the country's energy industry. It is significant to point out that many of the wells drilled over the last decade in the Inuvialuit Settlement Region have resulted from the acquisition of oil and gas rights that the Inuvialuit themselves have put up for sale on their own lands.

Divided Perspectives

Many Aboriginal leaders are key supporters of the Mackenzie Gas Project, arguing that oil and gas development is the only way Aboriginal communities—and the economy of the Northwest Territories as a whole—can achieve jobs and prosperity. Yet the project hearings offered the space for the expression of a diversity of views and perspectives that had not been heard previously. Beyond the rhetoric of northern leaders and politicians about economic opportunities, Aboriginal employment and the future of the NWT, the hearings revealed that there remain widespread concerns at the community level over the social, economic and environmental impacts of the Mackenzie Gas Project.

The support of Aboriginal political and business leaders has given industry, government and the media the impression of unequivocal support for the project, yet prior to the hearings the majority of Aboriginal voices had been muted. Until the start of the hearings, feelings of uncertainty over the project, as well as the extent of opposition to the pipeline, remained unknown. The hearings process gave Aboriginal peoples living in Mackenzie Delta and Valley communities an unprecedented opportunity to express their feelings, anxieties and concerns about the pipeline and facility operations. In Wrigley, southern NWT, D'Arcy Moses of the Pehdzeh First Nation gave voice to the concerns of many people who fear a loss of traditional culture in the face of energy development:

> *Our Elders speak of the fact that everything we need as people to maintain our way of life, and thus our identity is all around us. We are surrounded by clean water. We have an abundance of game. Our rivers and lakes teem with fish, and the land provides traditionally in all manner of plant material. It is these variables that are the core of our value system as a First Nation, and we ask: How is it that the external parties involved in the MGP can place a dollar value on this?* [24]

Ruby Koe spoke in Inuvik about the boom and bust nature of development and how communities are going to be hit hard by the MGP:

> *And you're just going to tear people's lives right apart and the land is going to be torn apart. It may be good for some people, but it's going to be bad for a lot of people because I've had family members that died with alcohol, and that's what we're already faced with. I have to take care of my kids. I have to prepare them for this, but they need to be educated, and they need to be told what's involved and how it is going to have an effect on them because, right now, they really don't know much about oil and gas. All it is is oil and gas. A lot of these elders here, all they hear is oil and gas. Nobody told them what oil and gas is about because they don't really know anything about that. So, it's just like saying somebody has an addiction, all these chemicals coming, it's the same way with the land*

and with humans. We're going to be faced with ruptured land and a ruptured life. So we have to think about what we're putting ourselves through. I know this is a big thing for the whole NWT, the Yukon, Alaska; it's a big thing. But how long is it going to last for one person to work for a certain amount of time? [25]

At the Fort Mcpherson hearings, testimony from Elaine Alexie of the Tetl'it Gwich'in Nation summed up the feelings of many young people who appeared at community hearings throughout the NWT:

I am opposed to the proposed Mackenzie Valley Pipeline. As a Gwich'in youth, I feel that this multi-billion dollar project will not only provide economic means to our communities, which is deemed through the eyes of the industry and of our own leadership as opportunity to our people, but I strongly feel that this development project will destructively affect and worsen the social, cultural spiritual, physical, and environmental well-being of our communities. The current state of the substance and physical abuse as seen now in the communities will worsen. With that comes our own loss of language and traditional culture, and in my own observation, this has already taken effect with our youth. There is nothing to safeguard the preservation of our traditional ways of life once the pipeline is built. We, as the people, have a right to clean air, human health, access to our environment, and most importantly, our food sources. The only way our culture is to survive is for us to secure our language, our spiritual and traditional beliefs and of a land that still maintains to sustain us. [26]

Kyla Ross, an 18-year old Gwich'in woman from Fort McPherson attending school in Lethbridge, southern Alberta, submitted written testimony in which she expressed her concerns about the pipeline:

This pipeline was stopped 30 years ago because our elders are wise. They knew what would become of it. This pipeline will be going through the Mackenzie Valley, through the regions of the

NWT and down to Alberta where the oil and gas will be shipped to other places worldwide. Sadly, we will only be profiting about 0.1 percent. Is 0.1 percent worth tearing up everything we have and leaving us empty-handed...Imperial Oil has blueprints on how the pipeline will go through, and we see that they are showing us that our land will not be disturbed. This is untrue. No construction company can put back what they take apart. [27]

Such sentiment is widespread, and it was articulated and supported by the Arctic Indigenous Youth Alliance (AIYA), a grassroots, non-profit youth organization based in the Northwest Territories, which had intervener status at the hearings. AIYA seeks to empower Aboriginal youth in northern Canada to engage with decision-makers in industry and government, and equip them with the information and knowledge they need to make decisions for a sustainable development framework based on the Dene and Inuvialuit traditions and culture. As AIYA stated in its original letter of submission to the Joint Review Panel:

We feel we are not given all the information to make an informed and balanced decision because Government and industry are fast-tracking and rushing the assessment of the project. [28]

At JRP hearings in Inuvik, Gerri Sharpe-Staples, President of the Status of Women Council for the NWT (which is an advisory agency to the Government of the Northwest Territories) spoke of concerns about the potential impacts of the project on community well-being. Specifically, she reported on the concern of northern women that communities were not prepared for the influx of outside workers:

In our opinion, and the opinion of many women, the induced effects will be long-term. The actual presence of thousands of southern workers may be short-term during construction only, but the related potential negative effects, such as teen births, HIV infection, increased drug use or increased family dysfunction are long term. If individuals are victimized through a project-induced in-

crease in family violence or sexual abuse, they will suffer long-term impacts. A large body of literature exists on the long-term impacts of violence and abuse and post-traumatic stress. Violence does not have to occur over a long-period of time for a victim to suffer long-term effects. As we know from historical abuse, such effect can also impact future generations. [29]

The Dehcho and the Dene Tha': Livelihood Rights

At the time of researching and writing this book, there were two major obstacles for the project proponents: the unresolved land claim of the Dehcho First Nations in the central Mackenzie Valley, and the legal action of the Dene Tha' of northern Alberta. The importance of these cases illustrates the importance of consultation and of existing treaty rights, as discussed in Chapter Three, as well as the cross-jurisdictional nature of megaprojects and the complexity of the regulatory processes and mechanisms currently in place.

The Dehcho First Nations of the Central Mackenzie Valley

The Dehcho First Nations is a tribal council representing 13 Dene and Métis communities in the central NWT, for whom a land claim settlement is an urgent priority. The proposed pipeline would run for approximately 40% of its length through Dehcho traditional territory. Although not opposed to the project, nor to membership of the Aboriginal Pipeline Group, for the Dehcho a land claim settlement is a precondition before discussions about their participation and involvement can begin. The Dehcho Declaration of Rights asserts that:

The Peace Treaties of 1899 and 1921 with the non-Dene recognize the inherent political rights and powers of the Deh Cho First Nation. Only sovereign peoples can make treaties with each other. Therefore our aboriginal rights and titles and oral treaties cannot be extinguished by any Euro-Canadian government. Our laws from the Creator do not allow us to cede, release, surrender or extinguish our inherent rights. The leadership of the Deh

Cho upholds the teaching of the Elders as the guiding principles of the Dene government now and in the future.

Today we reaffirm, assert and exercise our inherent rights and powers to govern ourselves as a nation.[30]

On 23 May 2001, the Dehcho First Nations signed two Agreements with the Governments of Canada and the NWT: 1) A Framework Agreement, which sets out the objectives, agenda of topics and negotiating principles of the treaty-making process, and 2) An Interim Measures Agreement which establishes the land-use principles and procedures that are to be observed during the several years it will take to negotiate and ratify a Final Agreement. These two agreements are the first steps towards a comprehensive agreement on outstanding land and self-government issues, which in effect will be a modern treaty between the Dehcho and Canada. The Dehcho emphasize that they have never surrendered title to their lands and territories and that treaties made with the Crown confirm they are the governing authorities on their lands.

The Dehcho argue that they are entitled to have revenue from the Mackenzie gas pipeline paid to them directly as a separate level of government. They are also asked for greater clarity around royalty sharing, better environmental assessment, greater understanding of the social impacts, information about impacts on caribou and moose populations and on traplines, and a guaranteed voice on the Joint Review Panel. The Deh Cho Interim Measures Agreement provides for participation in land and water regulation through membership of the Mackenzie Valley Environmental Impact Review Board and creation of a Dehcho panel of the Mackenzie Valley Land and Water Board. The Agreement also sets out the requirements for benefit plans related to oil and gas activities in the region, yet the Dehcho are critical of the MGP hearings process, as Chief Keyna Norwegian of the Liidlii Kue First Nation told the JRP:

With regards to this review process, we are concerned and disturbed by the decisions taken by the Joint Review Panel and others that have resulted in this process, becoming one that is clearly and significantly unfair and biased in favour of those who are supportive of the pipeline. [31]

The Dehcho have been criticized by the Aboriginal Pipeline Group for their position and have come under pressure to join the APG. In turn, the Dehcho have criticized the APG as being a partner with the energy companies only in the construction and operation of the pipeline, not as a partner that would own the gas that will flow through it. Suspicious that the energy companies are only using the Aboriginal Pipeline Group to help finance the construction of the pipeline, the Dehcho have agreed to consider joining the group only if they think it makes economic sense to do so.

At the beginning of the MGP hearings, Dehcho Grand Chief Herb Norwegian stated that there was no rush for them to join the APG. One of Norwegian's main concerns was over the economic viability of the pipeline and the rising and uncertain costs involved in the APG's participation in the venture, but Dehcho membership of the APG does raise problems as far as land claims negotiations are concerned. For many, contentious issues remain, such as a concern that Dehcho land would have to be developed to satisfy the deal made between the APG and the other project proponents. Dehcho members have also pointed out that the regulatory hearings process was based on the Mackenzie Valley Resource Management Act, which was created without consultation with the Dehcho First Nations. Leaders of the Dehcho First Nations state publicly that they believe that they signed a treaty with the Government of Canada in order to share the land with Canada. They did not release, cede or surrender the land, they argue. Some are also careful to point out that current negotiations with the Government of Canada should not be referred to as a land claims process as the Dehcho already have jurisdiction over their lands and that the existing treaty is an agreement to share that jurisdiction with Canada.

In October 2006, Alternatives North, a Yellowknife-based coalition of environmental NGOs and social justice groups, released a financial and economic assessment of the Mackenzie Gas Project. It shows that the project will generate huge revenues for the project proponents, while Aboriginal people and northern governments will benefit very little under the current royalty regime in the NWT. Alternatives North also critiqued the socio-economic agreement negotiated between the Government of the Northwest Territories and the MGP's proponents on the basis that there was no public involvement in its negotiation, drafting or review stages. Furthermore, the coalition was concerned that Aboriginal governments were not involved either and that discussion of socio-economic development had not been linked to a larger plan for sustainable development in the Northwest Territories.

The Dehcho position is that, as the current royalty regime will benefit energy companies and not Aboriginal and local people in the NWT, and as Canada shows no willingness to consider reforming it, then the Dehcho have to insist that Canada recognizes their jurisdiction over Dehcho lands and resources. Above all, the main worry for the Dehcho has been that, by joining the Aboriginal Pipeline Group, their negotiations with the federal government over a land claim deal would be impaired. While the Dehcho argue that the pipeline cannot be built without their approval and support, the Canadian federal Indian Affairs Minister Jim Prentice went on record as saying that the pipeline was crucial for economic development in the western Arctic and its construction would not be held up by the objections of one group (Weber 2006). Furthermore, an NWT territorial government report to the NEB stated that, taking into account the public interest in the project, it could not agree to a single community or region having a veto over approval of the MGP (Jaremko 2005).

The Dene Tha' of Northern Alberta
The final 103 km of the Mackenzie Valley pipeline will pass through northern Alberta and connect with an existing pipeline network operated by TransCanada Pipelines. There it will also link to the North Central Crossing Pipeline to Fort McMurray and the Alberta oilsands operations. This pipeline, currently being constructed by TransCanada's subsidiary, Nova Gas Transmission Ltd. (NGTL), is a connecting facility between the MGP and other infrastructure

currently in operation and was approved in October 2008 by the Alberta government. The pipeline will pass through the lands of the Lubicon Cree First Nation, which has outstanding land claims issues in the region (see Chapter Six). The Dene Tha' First Nation, representing 2,500 Aboriginal residents in northern Alberta, have argued that the Mackenzie Valley pipeline will not only pass through their territory but that their traditional lands reach into the Northwest Territories, where they overlap with Dehcho traditional territory. The Dene Tha' signed Treaty 8 in 1899 and the section of the pipeline extending south from the NWT border into northern Alberta would go through Dene Tha' territories as recognized and defined by that treaty. While the pipeline would not run through Dene Tha' reserves, it would cross traplines and hunting and fishing lands which remain of economic and cultural importance. Any connecting MGP facilities in northern Alberta are integral to the pipeline, they argue, and must be part of the federal review and discussion about them must respect existing treaty rights to hunt, trap, fish and gather plants for food as well as the duty to consult.

The Dene Tha' issue differs from the Dehcho situation but is based on a grievance that arises from similar concerns over decision-making and control of the Mackenzie Gas Project, exclusion from consultation and from the regulatory process, exclusion from the environmental assessment, and profound concerns over the social, economic and environmental impacts of the pipeline passing through traditional lands. The Dene Tha' are not necessarily hostile to the oil and gas industry—rather, they already participate in it and derive economic benefits from energy development. For example, they have a co-operation agreement with TransCanada Pipelines, they have a partnership in five drilling rigs with Calgary-based West Lakota Energy Services, and many community members work on oil and gas projects. The issue is one of consultation. The Dene Tha' argue that they have a constitutional right to be informed of the decisions being made that concern the MGP and its connection facilities. They claim they were not provided with an opportunity to have their opinions on the MGP heard, nor were they consulted by federal ministers despite it being their duty to do so and to accommodate Dene Tha' treaty rights.

The Mackenzie Gas Project is regulated by the NEB and is a federal government concern, but the difficulty for the Dene Tha' is that the Alberta section will be decided upon by Trans-Canada Pipelines and the Alberta Utilities Commission (AUC) and the Energy Resources Conservation Board (ERCB)—both were formed from the former Alberta Energy and Utilities Board (AEUB) in January 2008—making the regulatory process for this part of the route a joint concern for energy companies and the Alberta provincial government. Furthermore, First Nations communities in Alberta will not be included in federal community support programmes, socio-economic agreements, nor in industrial benefits deals with the energy companies. The Dene Tha' concern is that, as the final 103 km of pipeline and connecting facilities are merely defined as a routine extension of the existing TransCanada system, this crucial southern link is being disguised as a minor project by energy companies and by Alberta industry and pipeline regulators.

In May 2006, the Dene Tha' launched legal action against the project.[32] Their lawyers filed a judicial review with the Federal Court of Canada against the federal government, the NEB, Imperial Oil and the JRP, alleging that they had failed to consult with Dene Tha' leaders and communities and complaining that they had been left out of impact and benefit negotiations. They also maintained that their status in the regulatory review process—being only interveners—was inadequate, that the Alberta sections of the pipeline should be included in the federal review, and that the megaproject and its associated development infringed Dene Tha' Aboriginal rights and titles in NWT and Alberta. Earlier, in January 2006, they had requested that the JRP delay the hearings until after their applications for a judicial review. This request was denied, as the JRP said that many of the Dene Tha' concerns were beyond the scope of the regulatory review process.[33]

On 10 November 2006, a judgment was issued by Justice Michael Phelan of the Federal Court of Canada which prevented the JRP from considering, in the course of its hearings, evidence on matters involving the connecting facilities in northern Alberta or the territory in which the Dene Tha' First Nation have or have asserted Aboriginal or Treaty rights. In *Dene Tha' First Nation v. Canada*, the court concluded that federal ministers had breached their duty to consult with the Dene Tha' on the regulatory and environmental review processes related to the entire project, from its earliest inception to the present. Justice Phelan rejected the argument put forward by the defendants that no duty to consult had actually arisen.

Fogarassy (2007) points out the legal test to determine exactly when a duty to consult has arisen is set out in *Haida vs. Canada*, in which the Supreme Court of Canada points out that a duty to consult arises when a) the Crown has knowledge, real or constructive, of the potential existence of an Aboriginal right or title; and (b) the Crown contemplates conduct that might adversely affect such Aboriginal right or title. Referring to *Haida*, Phelan reasoned that the duty to consult arose with the creation of the Cooperation Plan for the MGP. The Cooperation Plan, he argued, was not merely conceptual in nature. It set out to do something, the objective being the construction of the Mackenzie Gas Project. It was a well-thought out roadmap to guide the environmental and regulatory review processes, from which the Dene Tha' were excluded. The Joint Review Panel was criticized for the one occasion it did decide to consult with the Dene Tha', which amounted to giving the First Nation a deadline of 24 hours to respond to a process which had taken several years to establish and which had already involved extensive consultation with everyone else who would be potentially affected by the MGP.

Late consultation is defined as inadequate consultation and if a court determines that consultation has commenced in a late manner, or not at all, then all Crown decisions or actions regarding a project are immediately suspect. A court determination based on inadequate consultation with Aboriginal peoples affected by a project could render a government decision on that project invalid (Fogarassy ibid.). Justice Phelan's ruling has significant implications for the legitimacy of the public hearings and for the Mackenzie Gas Project as a whole. The order had the effect of requiring the JRP to postpone several of its scheduled hearings. On 30 January 2007, Justice Phelan modified the original order to permit the JRP to address subject matters and complete the hearings that had been deferred. The JRP was further restrained by the court order for a while as the court also prohibited it from issuing its final report to the National Energy Board pending a later decision.

The NEB and JRP positions are clear: both assert their federal status and refuse to be drawn into a jurisdictional controversy with Alberta and TransCanada, while the JRP points out that it has no mandate or power to make Alberta enforce directives for wildlife conservation, habitat protection or community concerns. At the JRP hearings in High Level in northern Alberta, Dene Tha' leaders participated as interveners, turning them into a forum for spirited resistance. Chief James Ahnassay told the session, "We're participating under protest. We question the legitimacy of these hearings." He added that: "the process has become deeply hurtful and insulting to us".[34] For the Dene Tha', the hearings were an opportunity to relay to the panel the fact that they had not been properly consulted, that the oil and gas industry would adversely affect their use of the land, and they had not benefited from development in the past. Above all, elders reminded the JRP that oil and gas were finite resources, warning that the industry was merely a passing phase compared to the endurance of Aboriginal cultures. The Dene Tha' argument was also supported by some environmental groups such as the Sierra Club of Canada which, while pointing out the importance of the duty to consult and the accommodation of treaty rights, saw an opportunity to call for the NEB to regulate the entire pipeline project and to critique the Alberta Energy and Utilities Board as industry-friendly. In July 2007, it was announced that the Dene Tha' First Nation and the federal government had signed an agreement that would see the Dene Tha' receive Can$25 million to help address possible social and economic impacts resulting from the construction and operation of the Mackenzie gas pipeline.

Cumulative Impacts

A number of NGOs, both northern and southern-based, have also established a wide array of positions on the pipeline, arguing that the project has to be in Canada's interest as a whole. Many have expressed concern over environmental impacts and irreversible changes to northern ecosystems—for example, the Mackenzie Delta is of tremendous importance for some 175 species of birds, particularly the millions of seabirds and waterfowl that use the area for feeding, nesting or resting during migrations in the spring and autumn, while delta bays provide summer calving and feeding grounds for beluga whales. Environmentalists are also fearful of damage to caribou migratory routes nearby, and calving grounds in the Alaska-Yukon borderlands—the Porcupine caribou herd calves west of the delta, while the Bluenose caribou herd ranges to the east and south of the delta.

As the hearings for the Joint Review Panel began in February 2006, Aboriginal leaders in favour of energy development criticized environmental groups such as the Sierra Club and WWF for being similar to the anti-trapping organizations of the 1970s and 1980s that had impoverished many Aboriginal communities as a result of their successful campaigns against traditional resource-use activities. Yet the Sierra Club, an active intervener throughout the hearings process, claims that rather than sending a relatively clean energy source to replace the coal or diesel being burned in southern Canada and the U.S., the pipeline will carry gas from the Mackenzie Delta to northern Alberta, where it will be burned to heat up the viscous mix of bitumen, clay, sand and water that is becoming such an important cornerstone of Alberta's economy (a point—and a process—to which we will return in Chapter Six). The organization has also highlighted uncertainties and environmental risks involved in the construction and maintenance of the pipeline. It has also argued that the project should be seen in a wider context in terms of the environmental effects of hydrocarbon development, such as climate change, loss of biodiversity, permafrost degradation, negative effects on air and water quality, and the destruction of wilderness. Advocating that it is possible to balance industry, nature and culture, WWF-Canada has argued for the establishment of protected areas prior to development of an energy corridor, while the Canadian Arctic Resources Committee (CARC), also concerned that the pipeline will take fuel to the northern Alberta oilsands, carried out mapping projects of the cumulative effects of the pipeline.

The Alberta-based Pembina Institute, a research, education and advocacy organisation concerned with sustainable energy, continued to argue for consideration of the cumulative impacts of the Mackenzie Gas Project. Their report *A Peak into the Future* (Holroyd and Retzer 2005) claims that Northerners have been provided with little information that illustrates potential scenarios for oil and gas development in the Canadian North over a 30 to 50-year time period. Similarly, they argue, information about the potential cumulative, long-term ecological, economic and social impacts of full-scale natural gas exploration and development is limited. Holroyd and Retzer point out that the emphasis to date has been on individual gas projects, such as seismic projects, exploration drilling and the Mackenzie Gas Project, which represent only one stage of a much larger development process (the understanding of which was pointed out by Thomas Berger in the 1970s). Their findings suggest that Northerners can expect industrial development to increase significantly over a period of 10 to 20 years and then, unless more reserves are found, decline. Their report shows that the rate of development and ultimate environmental footprint will be similar to other mature gas fields in western Canada's sedimentary basin that, following a boom in northern Alberta and northern British Columbia, are now fully developed and have left significant surface disturbance on the landscape.

Holroyd and Retzer's concerns over the lack of consideration of cumulative impacts are significant given that, at the time they wrote their report, exploration activity for oil and gas had already stepped up in anticipation of the Mackenzie Valley pipeline decision. And while discussion of cumulative impacts of energy development was of central concern to Berger's inquiry, the MGP hearings generally played down this aspect of the project, with the proponents only providing some information when pressed, and arguing that discussion of future development beyond the MGP was not within the scope of the MGP regulatory review.

When the JRP held its community hearing in Ulukhaktok on Victoria Island, north of the Canadian mainland in September 2006, the community that would be furthest away geographically from any of the MGP components and infrastructure, local people expressed concern at the sidelines and during the coffee breaks about Ulukhaktok becoming a base for oil and gas exploration. People asked each other, "Why is this panel really here?" One of the community's high school teachers told the panel that he had reflected on the route and magnitude of the project:

And it suddenly struck me, looking at that, that we weren't just talking about one pipeline going 1300 kilometres; we were talking also about what that would open up once the pipeline was in place, and it – you know, all of the area to the east and west of that pipeline, the whole territory. It seemed to me that then you would be talking about the possibility of other pipelines being built and then connecting into that central one, so that it wasn't just an issue of this thin line down the middle, it was an issue of the entire territory then branching out from that so that you could conceivably see the whole territory covered in a network of pipelines eventually. And nobody really seemed to address that issue, the issue of what happened later once that pipeline is in place, because it seems like a pretty good opportunity once you've got it there to then have other companies come in and say: Well, we can use that – we can develop all the rest of the territory and then just tie into that main pipeline. [35]

Local resident and elder Alberta Elias asked a direct and pointed question about the cumulative impacts to the proponents:

My question…is directed to the proponents. If the pipeline was built, isn't it the case that the Beaufort Sea and other coastal areas near Ulukhaktok would be open for exploration and development? And what kind of social and environmental impacts could there be from these developments? And the third part of it is this: Isn't it true that one of the reasons for building – the idea of building the pipeline is to develop more areas other than the three initial sites? And this is all about cumulative impacts. [36]

The response to this question came from Randy Ottenbreit of Imperial Oil and his reply is worth quoting in full:

In the...in sizing how much natural gas the gathering system and the Mackenzie Valley Pipeline might be able to ship, we have designed it in a way that provides extra space or capacity for other gas to be shipped. And our expectation is that once a pipeline is built, that it will provide some encouragement for people to explore for natural gas and to develop it, should the exploration be successful. We don't know how successful that exploration will be and how much development would occur. We also don't know where future discoveries would be made. So because we don't have that kind of information, we've focused on the assessment of the project that we do know, and that's the Mackenzie Gas Project. And there has been some work done at the Panel's request – the request of this Panel around what's called some hypothetical scenarios, but we don't know what the future success for exploration programs would be, so it's tough to define what future impacts would be. With respect to kind of offshore exploration, that exploration could continue independent of whether the Mackenzie Gas Project proceeds or not. There's a process in place for the licensing – the issuance of exploration licenses. As you're aware, there was a fair bit of exploration in the Beaufort Sea in the 1980s, and that did not result in any subsequent development. There were some discoveries made, but it has not been developed. The Mackenzie Gas Project, that we've defined, focuses on the development of onshore natural gas as opposed to offshore natural gas. That doesn't mean that offshore natural gas that has been discovered might not be developed at some point in time in the future, but it will be the subject of its own proposal and regulatory review process and involves a different kind of development than onshore natural gas fields. So if I were to summarize it, yes, if the Mackenzie Gas Project proceeds, it will likely encourage people to explore for more natural gas and develop it, if they're successful. But we don't know the extent to which they'd be successful and when development would occur. Should they be successful, should there be additional development, it will be the subject of its own regulatory review, as well. But in the meantime, we have, at the request of the Joint Review Panel, provided information around what's called some hypothetical development scenarios, and that is available and it's publicly available information.[37]

The Vice-president of Markets for the Canadian Association of Petroleum Producers has said that the Mackenzie Valley pipeline proposal is the real driver of all Arctic oil and gas exploration, whether on land or in the Beaufort Sea, confirming that the energy companies operating in Canada's North will focus their activities on exploration in the mid-Northwest Territories, the Mackenzie Delta and the Beaufort Sea.[38] The major energy companies are positioning themselves and used the hearings to ensure that the National Energy Board recommends that the pipeline will have "open access" to anyone who wants to bring gas into the system, as well as addressing the issue of the cost of service tolls for transporting gas, which are directly related to the capital costs of the infrastructure. The higher the cost of the pipeline, the higher the tolls based on a percentage.

Pipe Dream?

The Mackenzie Gas Project illustrates that renewed interest in developing the oil and gas resources of northern Canada presents Aboriginal peoples with possible economic opportunities for developing sustainable livelihoods and communities as well as with significant social, cultural and environmental risks and impacts. From Inuvik in the northern NWT to High Level in northern Alberta, and from the small communities of Sachs Harbour and Ulukhaktok on Canada's Arctic islands to Alberta's provincial capital of Edmonton, the MGP hearings were filled with rich testimony from Aboriginal people about both the memories and the present realities of traditional hunting, fishing and trapping ways of life, and of the contemporary social and economic situations of northern communities. As the hearings and their recorded transcripts show, people have spoken powerfully and emotionally about being out on the land, but they have also expressed their fears for the future of Aboriginal communities both with and without the oil and gas industry. As was evident during the Berger Inquiry, the MGP hearings also affirmed how stories of traditional life and testimony about the persistence of Aboriginal culture also provide a discursive context for the expression of hopes for the achievement of economic independence and cultural survival.

For Aboriginal peoples—and indeed all residents of the NWT—the hearings offered the opportunity and space for open conversation and debate (although this was constrained somewhat compared to the hearings Thomas Berger conducted in the 1970s), for the exchange of information and ideas, and for a greater understanding of the scope of the project before a final decision is made and the specific conditions are set out. Yet the Dene Tha' case reminds corporations, governments and Aboriginal business leaders of certain rights and duties, and of the recognition and affirmation of existing Aboriginal livelihood and treaty rights in Canada's *Constitution Act, 1982*. Put simply, Aboriginal peoples in Canada inherit ancestral rights based on their historic occupation and use of traditional territories and resources. Project proponents have a duty to consult Aboriginal communities to be affected by those projects, as we saw in Chapter Three when I discussed the historical, cultural and political importance of treaty rights and treaty commitments.

With other large-scale development proposals submitted and pending in northern Canada, the Mackenzie Gas Project hearings have demonstrated that government and industry need regulatory certainty before making informed economic decisions, as well as an understanding of the social, cultural, political, legal and economic situations of indigenous communities, that Aboriginal people need access to reliable and detailed information, and, above all, that industrial projects in native areas can only proceed effectively if Aboriginal communities are properly consulted. In response to the release of the JRP's report endorsing the Mackenzie Gas Project, WWF-Canada welcomed the panel's 176 recommendations as "a necessary antidote to the Proponents' unhelpful attempts to limit the scope of discourse and minimize its obligations, an approach it has now extended to its response to the Panel's recommendations". WWF-Canada's reference was to the attempt by industry to request that the NEB should ignore most of the JRP's recommendations. Grant (ibid.) has argued that this is a "violation of the process". In response to some of Imperial Oil's comments on the JRP report, the Dehcho First Nations have suggested that

> *Imperial seems particularly troubled by recommendations that require it to obtain community approval. Imperial prefers instead that it just be required to consult and that the NEB decide on matters of disagreement between Imperial and communities. In making this argument, Imperial clearly does not recognize or accept any right of communities to make their own decisions on their traditional lands, or for Imperial to require community approval, based upon aboriginal and treaty rights. Obviously, we do not accept Imperial's perspective on this issue.* [39]

The Dehcho First Nations submitted their response to the JRP report and reminded the NEB, the Government of Canada, and the Government of the Northwest Territories that "the assessment, mitigation and accommodation of infringements upon Aboriginal and Treaty rights by the MGP was not part of the JRP's jurisdiction". Furthermore, they argued that consultation between the Government of Canada and the Dehcho Aboriginal governments would still be required, regardless of what decision was made about the MGP nor, they pointed out, should any of the JRP's recommendations "dictate or prejudge the scope of issues or remedies to be discussed in those consultations. There has not yet been any substantial attempt to consult, or to even discuss a consultative process, by the Government of Canada on potential infringements of Dehcho Aboriginal and Treaty rights by the MGP." Furthermore, the response pointed out,

There are still numerous site-specific Dehcho concerns about this project that have not yet been addressed. The information provided to the JRP in many instances was at a high level and so lacked the details necessary to engage in a discussion of specific Dehcho concerns. As a result, many of the JRP's recommendations are necessarily at a high level as well. While some of the JRP recommendations can create a framework upon which the site-specific concerns can be discussed and possibly dealt with, discussions on these specific concerns still need to occur and issues have to be resolved.[40]

In its submission to the NEB in response to the JRP report, the Sambaa K'e Dene Band claimed that the report

falls well short of fulfilling Canada's consultation/accommodation obligations for two reasons. First, the JRP report primarily speaks to the MGP at a broad level and establishes a project-wide "framework" for environmental mitigation rather than articulating specific mitigation measures. This approach is partially due to the scope of the project but also, as noted by the JRP, is due to the fact that the Proponents provided very general information in its EIS rather than the type of detailed information that would lead to detailed mitigation recommendations. Most of the recommendations made by the JRP require considerable more research, planning, and mitigation decision-making by a wide range of stakeholders. In most cases, the JRP simply did not have the type of information required to address area and project-specific mitigation, which has been SKDB's focus since the beginning of its interventions. This lack of detail limits the usefulness of the report's recommendations with respect to SKDB area-specific concerns.

Second, as discussed further in the section on Chapter 5 below, it is SKDB's view that both the Proponents and the JRP have either underestimated or overlooked the significance of MGP impacts within the SKDB traditional land use area, due to differing perspectives on the project. This underestimation of environmental impacts, which might lead to an underestimation of the nature and degree of rights infringements, along with avoidance of address-

ing s.35 issues directly, means that a few of the key concerns and potential accommodation measures proposed by SKDB have been avoided or ignored by the JRP and the Proponents. [41]

The importance of consultation is set out in the United Nations Declaration on the Rights of Indigenous Peoples (UNDRIP), which was adopted by the UN General Assembly on 13 September 2007. Canada finally endorsed UNDRIP in November 2010. As pointed out in the introductory chapter of this book, Article 18 of UNDRIP affirms that indigenous peoples have the right to participate in decision-making in those matters which would affect their rights, while Article 32 places the responsibility on states to consult and cooperate in good faith with indigenous peoples in order to obtain their free and informed consent prior to the approval of any project that affects their lands or territories and other resources, particularly if the project intends to develop, utilize or exploit mineral, water or other resources. In proclaiming the UNDRIP, the UN General Assembly also considered that treaties, agreements and other constructive arrangements are the basis for a strengthened partnership between indigenous peoples and states. While indigenous peoples claim that UNDRIP should be seen as part of customary international law, Canada's initial refusal to ratify it was based partly on a view that it is not a legally-binding instrument.

Consultation aside, however, the project's future may well be decided not by the National Energy Board or the Government of Canada but by the proponents themselves, based on rising costs and impatience with the delay in the regulatory process. The cost estimate for the project in October 2004, when the original application was submitted, was Can$7.5 billion. In March 2007, Imperial Oil (which is approx. 70% owned by ExxonMobil) raised this estimate to Can$16.2 billion and filed a letter with the JRP that provided an updated cost estimate and schedule for the MGP, as well as associated adjustments and refinements to the project. At its annual shareholders meeting in Dallas at the end of May 2007, ExxonMobil's chief executive told the media that, at such increased costs, "it's not viable to build that pipeline.....It may just be that this project will have to wait for a different cost environment."[42]

In October 2009, Canadian newspapers, citing unconfirmed government sources, reported that the Canadian government was, in all likelihood, going to renege on a promise to provide subsidies and fiscal incentives to the proponents. Imperial Oil has since indicated to the NEB that, if approval is given for the project to go ahead, construction on the pipeline will probably not begin until 2016, with gas beginning to flow in 2018 at the earliest. Supplies of gas from rival projects, such as shale gas development (shale gas is an unconventional but increasingly important source of natural gas in North America), may well render the Mackenzie Gas Project uneconomic but the delay to the MGP, as well as escalating costs, also raise the prospect of the Alaska Highway Gas Pipeline emerging as the most effective and realistic option for transporting Arctic gas to the south. This is the subject of the next chapter.

CHAPTER V

THE ALASKA HIGHWAY GAS PIPELINE AND THE LAST WILDERNESS

The original idea for production of Arctic gas was to bring it in a single pipeline from Alaska across the North Slope and the coast of the Arctic National Wildlife Refuge, along the top of the international boundary between Alaska and Yukon and then, with gas from the Mackenzie Delta piggybacking, continue through the Mackenzie Valley. Thomas Berger's concerns about the level of ecosystem disturbance this project would bring to the Arctic coast led to the northern route from Alaska to Canada being shelved. While dreams of constructing a pipeline along the Mackenzie Valley route were kept alive by industry and governments until the political, economic and regulatory environments were considered to be favourable, the creation of Ivvavik National Park in 1984 meant the prospect of an energy corridor ever being established from Alaska along the northern coast of Yukon would remain a distant one. The park was created to set aside and protect the ecological integrity and wilderness characteristics of the northern Yukon and western part of the Mackenzie Delta.

Yet although the first Mackenzie Valley gas pipeline proposal put forward by the Canadian Arctic Gas consortium was not approved by the National Energy Board, the NEB did endorse the alternative Alaska Highway pipeline proposal. Put forward in 1976 by Foothills Pipe Lines (now owned by TransCanada), the plan for the Alaska Highway Gas Transportation System (ANGTS), as it was formally called, was for the development of a pipeline along a route that would take gas from northern Alaska to U.S. markets by following the Trans-Alaska Pipeline System (TAPS) to Fairbanks (see Coates 1991, for a comprehensive discussion of TAPS), then travel along the route of the Alaska Highway to Canada, moving south-west through British Columbia into Alberta and then to the border between Saskatchewan and the United States. Decisions were made and approvals were granted and the ANGTS process seemed far more straightforward than the Mackenzie Valley proposal. The U.S. Congress approved construction of the pipeline, selected the route and established a statutory framework for its construction and operation under the Alaska Natural Gas Transportation Act of 1976. Part of the original Foothills proposal involved the addition of a pipeline from the Mackenzie Delta south through Yukon along the route of the Dempster Highway (a road between the Klondike Highway in central Yukon and Inuvik in the Northwest Territories, envisioned by John Diefenbaker in the 1950s and eventually opened in 1979), where it would connect with the Alaska Highway pipeline.

As it would follow established rights-of-way (i.e. an existing pipeline and road system) and, it was suggested, not disturb wildlife and wilderness in the way that it was feared the Arctic coast route would do, the Alaska Highway proposal was argued by proponents and regulators as being deemed less controversial to concerned Canadian indigenous groups and environmentalist organizations. For one thing, as far as the supporters of the pipeline saw it, there were no important calving grounds for caribou herds in the way, or migratory routes which were likely to be disturbed, and it would follow the Alaska Highway for much of its way south. The precedent that Berger had set with the Mackenzie Valley inquiry, however, forced the Canadian federal government to launch two inquiries, one on the environmental consequences of the route, which was carried out by the Environmental Review Office of the Department of Fisheries and the Environment, and a second on the socio-economic impacts on the Yukon, known as the Alaska Highway Pipeline Inquiry, chaired by Kenneth Lysyk, a dean of law from the University of British Columbia in Vancouver.

The environmental assessment concluded that the environmental impacts would be minimal, while Lysyk's conclusions were similar to Berger's. In his 1977 report, he recommended against pipeline development at the time, but he was more favourable to the idea of a pipeline through the Yukon than Berger was in his assessment of the Mackenzie Valley pipeline. Lysyk drew attention to the importance of considering the social, economic and environmental changes that would inevitably follow the construction of the pipeline. He also suggested that there may be later support for the project conditional on land claims being settled and if energy companies agreed to pay to mitigate the social and economic impacts, which he also believed could be kept within acceptable limits. Lysyk suggested a delay of four years to allow for negotiation and settlement of land claims. Ottawa passed the Northern Pipeline Act in 1978—effectively signalling an agreement between Canada and the United States on the go-ahead for the project—which led to the creation of the Northern Pipeline Agency. This was given the responsibility of overseeing the construction of the pipeline and establishing the terms and necessary conditions to deal with its socio-economic and environmental impacts (Dacks 1981). One of the stated objectives of the Northern Pipeline Act is to facilitate efficient and expeditious planning and construction of the Alaska Highway pipeline and take into account local and regional interests, the interests of local residents affected by the project, particularly indigenous populations, and to ensure that any Aboriginal claim related to the lands on which the pipeline is to be built is dealt with in a just and equitable manner.

Because it was not mired in land claims negotiations, construction on the Alberta section of the pipeline went ahead and was completed in 1982. It remains in operation and delivers something like one-third of Canada's gas exports to the U.S., sending natural gas mainly to the west coast and Midwest.[43] Yet overall, there seemed to be little in the way of the Alaska Highway pipeline project. Hearings were held, there was an environmental assessment and review process, certificates of public necessity and convenience were issued, and the pipeline right-of-way was established. All seemed to be in place apart from land claims settlements with Yukon First Nations. In the end, however, Foothills and the major natural gas producers with stakes in the pipeline did not feel the economic conditions justified the completion of the Alaska Highway pipeline project and it was not until 2001, with rising demand—along with sustained prices—for natural gas, that North American Arctic gas producers decided to move ahead with a new application to construct the Alaska Highway Gas Pipeline, or AHGP (Roddick 2006), which is also referred to as the Alaska Highway Pipeline Project, or AHPP.

The Alaska Highway Gas Pipeline: Current Status

In the United States, both Houses of Congress approved legislation in 2002 and again in 2003 to encourage the construction of the pipeline. In the middle of the decade, Coates and Morrison (2005: 310) wrote that "the pipeline, like earlier visions of megaproject salvation, remains in the future, still only a vision of resource-based prosperity". Yet it appears increasingly to be a real possibility that it may be built. In 2007 Sarah Palin, in her capacity as Governor of Alaska, invited industry to submit pipeline project applications within the framework of the Alaska Gasline Inducement Act (AGIA). In August 2008, Foothills (TransCanada) received an AGIA license, which has granted the company preferred status in Alaska. The Foothills project is estimated at around US$26 billion. In June 2009 it joined forces with ExxonMobil. ConocoPhillips and BP are also competing to build the pipeline under the banner of the Denali Project partnership, at a higher cost of some US$32 billion, and are doing so outside the AGIA, although these companies all have different views about financial incentives, such as risk-mitigation tax credit and subsidy packages, and the nature of government support needed before they would actually support the pipeline.

In addition to the Alaska Highway pipeline proposal, an alternative pipeline route entirely within Alaska is also under consideration. This would carry North Slope natural gas from north to south through Alaska to the port of Valdez in the same way that oil is transported from Prudhoe Bay. At Valdez, the gas would undergo liquefaction to produce liquefied natural gas (LNG), which would be transported by tankers to southern markets, many of them likely to be in Asia. Currently, the Alaska Highway route is the most likely option, although the main issues relate to the earlier decisions made about the project and the contemporary validity of the approval obtained in the late 1970s. For example, uncertainties remain as to the validity of the rights-of-way granted to the original proposal and whether there are additional possibilities and opportunities for interested parties or organizations to participate in the development of the current project (e.g. see Cowling 2001). The environmental assessment carried out at the time of the original proposal looked at the potential impacts on the land, but these were primarily federal lands which were administered by the Minister of Indian and Northern Affairs pursuant to the *Territorial Lands Act* (Cowling ibid.: 85). The current proposal may have to go through a process of applying for additional approvals because much of the route will cross First Nation lands, but also because of the requirements of the *Umbrella Final Agreement* (finalized in 1990, this established a framework for a general agreement on land claims and came into force in 1993; see below) as well as the Northern Pipeline Act. Yukon First Nations did not begin to settle land claims until the 1990s and, unlike in many other parts of Canada, did not negotiate and sign treaties with the Crown. As Cowling (ibid.: 88) puts it, the issue of land claims requires consideration of whether the Alaska Highway Gas Pipeline can be constructed prior to the settlement, between the Government of Canada and Yukon First Nations, of remaining outstanding land claims:

An example of such a possible conflict can be found in the approximately 1700 km of ANGTS that is proposed to be constructed through the Kaska's traditional territory. Although, the resolution of such an issue does not directly involve ANGTS, some First Nations may argue that resolution of the Yukon land claims agreements is necessary before ANGTS can be constructed. This implies potential delays in finalizing ANGTS. As a practical matter of law, the issue might be framed in terms of the ability of First Nations to secure and maintain an injunction against the project until land claims are settled.

For Yukon First Nations who have been watching the Mackenzie Gas Project process play out, the Dene Tha' case in northern Alberta has acquired particular salience. From the perspective of the Alaska Highway Aboriginal Pipeline Coalition (AHAPC—see below), the Supreme Court decision which ruled in favour of the Dene Tha' was also a positive one for Yukon in that it serves the AHAPC in its mandate to be proactive in preparing First Nations for the proposed gas pipeline project along the Alaska Highway. Specifically, given that there remains uncertainty over whether the regulatory authority will be under the auspices of the National Energy Board or National Pipeline Act, the Dene Tha' case is seen to provide a way of thinking about how Yukon First Nations will engage with governments and industry in the regulatory process. For the AHAPC, the judgement "makes clear the obligation of the Crown to consult with First Nations on matters that may affect their Aboriginal rights. It also provides a more clear understanding of how the Crown has operated in the Mackenzie for consideration in the development of a regulatory process that will provide for meaningful consultation of First Nations" (Alaska Highway Aboriginal Pipeline Coalition 2006: 1).

Estimated to take at least seven years to complete, the AHGP will be a massive construction project cutting across the north-west of North America. In anticipation of these developments, the Athabaskan peoples of the southern Yukon Territory, whose traditional territories the pipeline will pass through or near to, have begun to prepare for the possibility of pipeline construction, with all the attendant, negative social impacts feared to follow in its wake (Roddick ibid.). In particular, lessons learned from the regulatory process and public hearings of the original proposal, as well as the experiences of indigenous peoples with oil and gas development elsewhere in the North (including Alaska, the Northwest Territories and northern Alberta) and the experiences of Aboriginal people with previous major project developments in the Yukon, are seen as crucial to helping Yukon First Nations engage with industry and pipeline companies in a meaningful way as they face the Alaska Highway Gas Project in its current manifestation.[44]

In 2002 the Kwanlin Dun First Nation, which is located mainly in and around Whitehorse, produced a *Pipeline Engagement Study*, and the Kaska Tribal Council produced their report, the *Alaska Highway Natural Gas Pipeline Proposal*. Both documents echoed the Lysyk report with recommendations that included the need for pipeline companies and contractors to demonstrate how they would work with First Nations, and called for improved access to education, training and skills. The First Nations' reports emphasized the need to recognize the health, social and cultural issues that are connected to pipeline development .

With the encouragement of industry and government, Yukon First Nations along the proposed pipeline corridor decided to take a leadership role and formed the Alaska Highway Aboriginal Pipeline Coalition (AHAPC). Based in Whitehorse, the AHAPC is a registered society that acts as the central coordinating body for sharing information about the project between First Nations, industry and government. Leaders from First Nations communities identified a need to take a proactive approach and to be organized to the extent that they would be in a strong position to evaluate the nature of the project, develop programmes that would benefit First Nations communities and position themselves to be able to participate in the project during planning, construction, operation and remediation stages. The AHAPC sees itself as having an important role to play in advising government and industry about First Nations interests, advocating on behalf of the collective interest of First Nations, and ensuring assessment processes address the concerns and priorities of First Nations. Currently, the AHAPC is a coalition of five Yukon First Nations—Kwanlin Dun, Kluane, Carcross Tagish, Champagne and Aishihik, and Ta án Kwach an Council. Four other First Nations are observers. A Yukon First Nation whose traditional territory will be crossed by the right-of-way proposed for the pipeline is eligible to apply for membership.[45]

With the approach it takes as a resource centre, with no mandate or claim to negotiate with industry on behalf of First Nations, there is a clear difference between the AHAPC and the Aboriginal Pipeline Group in the Northwest Territories. The AHAPC remains neutral with respect to the pipeline project, including taking a position on what kinds of partnerships, if any, could be formed with the pipeline proponents, whereas the APG was formed to achieve an equity partnership role in the construction and operation of the Mackenzie Gas Project (Roddick ibid.). The AHAPC argues that its aim is not to take away or diminish the rights of any of Yukon's First Nations to make any decisions about the Alaska Highway Gas Pipeline project. It acknowledges that each First Nation in Yukon must maintain its right to consult and negotiate with industry and government, either alone or together with one or more of the other First Nations, and can decide on how they want to handle discussions over their own impact and benefit agreements. Industry and government must also be prepared to consult individually with each First Nation.

While First Nations are preparing to face the inevitability of oil and gas development in the Yukon, within the territory a number of unanswered questions remain that say something about the feelings of uncertainty many Aboriginal people have about the project, such as the nature of the assessment process, and what the regulatory process will look like, what form the tax structure will take, and whether there will be toll fees and a royalty structure. Furthermore, Yukoners are asking if they will have access to the gas that will flow through the territory, and they have concerns about equity opportunities and who is going to fund First Nations to be able to participate in the review process. Despite these and many more questions, the AHAPC aims to be well prepared to facilitate the consultation process and to be in a position where it can suggest standard approaches and encourage best practice where possible. The primary focus of the AHAPC is on four areas of common interest to Yukon First Nations: 1) socio-economic impacts of the proposed pipeline; 2) impact benefit agreements; 3) regulatory issues; and 4) environmental issues. The coalition points to the urgency of identifying gaps in existing environmental data, the importance of working with industry and government to fill those gaps, and the fundamental requirement of considering traditional knowledge and community land-use information in all environmental decisions.

While the APG aims to provide a model for advancing corporate-Aboriginal economic relationships and business interests, the AHAPC is more concerned with advocating new models for adaptive environmental and socio-economic assessment and management. It is consistent with the way Yukon First Nations have advocated for self-government as a way of ensuring they have "a substantive role in the management of the lands and resources in the Yukon and, perhaps more importantly, the power to provide effective and fair governance for their communities and citizens" (Leas 2005: 118). As Roddick (ibid.) has reported, the central concern is the boom and bust nature of development.

Approximately 26% of the 33,000 people living in Yukon are Aboriginal and there are 14 First Nations groups. The Yukon has four levels of government: federal, territorial, First Nation and municipal. The federal government has ownership and control of the territory's public land, water and resources. First Nation land claims settlement negotiations have been ongoing since 1973 and the Yukon territorial government has been negotiating the transfer of federal programmes to local and regional control. An example relevant to oil and gas development is the devolution to the Yukon territorial government and First Nations of responsibility for environmental assessment under the *Yukon Environmental and Socio-Economic Assessment Act* (YESAA). Under YESAA, development assessment legislation is a process required of every Yukon First Nation Final Agreement and ensures that the concerns and aspirations of Aboriginal people are recognized in the assessment of development projects. Specifically, First Nations can use YESAA in their negotiations with industry and government to make them aware of Aboriginal concerns about social, cultural and health issues. The *Umbrella Final Agreement* provided a framework for Yukon First Nations' final and self-government agreements, which have subsequently been realized for 11 First Nation governments that now operate as self-governing jurisdictions under the federal *Yukon First Nation Self-Government Act* (1995). They have responsibility for the administration of land claims rights and benefits (Leas ibid., Roddick ibid.). Another three Yukon First Nations, still negotiating their land claim settlements, operate as band councils under the federal *Indian Act*. Most Yukon First Nation governments also participate in one or more regional tribal organization. The largest regional body, the Council of Yukon First Nations, represents nine self-governing Yukon First Nations. The Kaska Tribal Council represents five member governments in south-eastern Yukon and

British Columbia, and the Gwich'in Tribal Council represents four communities in northern Yukon and the Mackenzie River Delta area of the Northwest Territories.

Should it proceed, the Alaska Highway Gas Pipeline may well prove to be the largest private-sector enterprise project ever undertaken in North America. Historically, Aboriginal people in the Yukon have not seen too many benefits from major development projects that have taken place on their lands. These projects have been transformative and, in some cases, have had dramatic impacts on family life, social relations and traditional land-based harvesting activities. The cumulative and longitudinal effects of these projects, but also the persistence of memories and narratives about them, have also helped to shape the ways in which indigenous people in the Yukon think about, reflect upon and respond to current and planned development, such as the Alaska Highway gas pipeline. With knowledge of the negative impacts of the past, they have different kinds of expectations about how development should proceed and how they should benefit from it (Roddick ibid.).

The 1896-98 Klondike Gold Rush brought some 40,000 prospectors and other travellers and wanderers into the Yukon. Some Aboriginal people found themselves gainfully employed as packers, guides, deckhands on sternwheelers, and as hunters providing food for the incomers. But indigenous trade networks were disrupted and the health impacts were often severe. Above all, a new territory was established in 1898 that required new forms of administrative and legal control, speedily ushered in from the federal government in Ottawa. In 1942-43, construction of the Alaska Highway through southern Yukon and continuing to Fairbanks in Alaska and the Canol pipeline from Norman Wells to Whitehorse brought in tens of thousands of construction workers, both from the U.S. military and civilian contractors. As well as the social problems this caused, U.S. soldiers seriously depleted wildlife through overhunting and this led to the creation of the Kluane Game Sanctuary. Cruikshank (2005: 66) describes how this acted to protect and conserve wildlife and banned all hunting within the sanctuary, including the traditional activities carried out by Aboriginal people who were forced to relocate to new but less productive hunting grounds on the margins east of the Alaska Highway. Coates and Morrison (2005: 241) describe how this constituted an invasion that "affected all aspects of Yukon life, created the greatest boom since the Klondike gold rush, and set the territory on a markedly different course".

Between 1974 and 1977, Alaska Natives witnessed a similar influx of migrant labour with construction of the Trans-Alaska pipeline. This project, along with the Yukon Gold Rush and Alaska Highway construction, also continues to stand as an example of what happens to places and communities which experience rapid in-migration from transient workers. The past experience of massive in-migration in Yukon has also shown that the negative social effects may persist for generations after the sojourners have left (Coates 1985, Roddick ibid.). Narratives of this experience emphasize the continued importance of land and language for Aboriginal identities and, in talking about how to be prepared for pipeline development, older people often reflect about growing up on the land and the importance of speaking in the words of one's traditional language. They are powerful narratives, speaking of past times and of the procurement of meat and fish, of how in spring and summer meat and fish were preserved for use in the winter, how berries and plants were gathered and cached in the ground for winter use. Skins from animals such as caribou were used for clothing, shelter and bedding. Nothing would be wasted. The land nourished the people who, in turn, looked after it.

Today, Yukon First Nations continue to speak for the land and are concerned about its future. Yukon First Nations governments have a strong say as to how land should be used and managed and they argue that they have an obligation to ensure that Aboriginal people receive a lasting share of any benefits that may result from a pipeline constructed across their lands. Through their self-government agreements, many Yukon First Nations have also strengthened their abilities and rights to participate in negotiation with industry, but these rights are also guaranteed under federal legislation and, arguably, under the Canadian constitution. This is a vastly different situation from the one Aboriginal people found themselves in during the time of the Alaska Highway Pipeline Inquiry. There is a critical role that First Nations argue they should play in environmental assessment processes and other land management decision bodies that are required to give the approval for elements for the project to go forward. Another important consideration is the First Nation communities that do not fall directly along the pipeline corridor but will ultimately be impacted if it is constructed. The AHAPC is careful to point out that, although it has a mandate to work for the interests of First Nations in Yukon, nothing displaces the power and authority of those First Nations. There is no transfer of governance rights to AHAPC. It is a coalition concerned with ensuring that First Nations have both a contribution to make to development plans and an economic gain from any development that takes place on their lands. The AHAPC's concern is that the construction and operation of the Alaska Highway Gas Pipeline does not compromise the integrity of the environment, lands and resources of the traditional territories of Yukon First Nations, but that it nonetheless brings economic and business opportunities and benefits. There is awareness that the project will have long-term impacts on First Nations' lands and peoples, so the aim is to advocate for long-term financial benefits that will flow to Aboriginal communities. This might include provisions for part ownership or some other business arrangement of long-term benefit to local communities.

The Last Wilderness: People, Caribou and Alaska's Arctic National Wildlife Refuge

Oil began flowing through the 800-mile-long Trans-Alaska Pipeline from the North Slope in 1977. Yet the petroleum industry has been the most important resource in Alaska since the discovery of oil at Prudhoe Bay in 1968. According to official U.S. government statistics, the Prudhoe Bay oil fields produce about 264,000 barrels per day and Alaska's total oil production provides just over 13% of the U.S. domestic supply of energy.[46] Revenues continue to supply about 85% of the Alaskan state budget and, because of its reliance on these revenues, it has been argued that Alaska quickly emerged in the 1970s and 1980s as a petro-state (Pretes 1991). Karl (1997) has suggested that, in a petro-state, the economy and politics are shaped by the influx of petrodollars to such a great extent that it is set apart from other states. For one thing, petro-states are far more dependent on a single commodity than other states are—for example, mineral-producing states—and the industrial sectors of the economy are closely linked to the development and production of the primary commodity. Although Karl focused on developing counties such as Venezuela, Nigeria, Algeria and Indonesia, Pretes (ibid.) has argued that Alaska has more in common with other oil-producing nations than perhaps other oil-producing states in the United States. Indeed, the emergence of oil as the predominant commodity in the state's economy led Tussing (1984) to describe how Alaska resembled Kuwait and Libya more than it did Texas or Oklahoma.

In northern Alaska, oil has also transformed the social, cultural and economic landscape of much of the region within the borders of the North Slope Borough, which is home to some 7,400 people, the majority of whom are Inupiat Eskimos. While there have been many benefits to Alaska Native communities in northern Alaska, including jobs, investment in schools and improved medical care, oil infrastructure and development have nonetheless had significant environmental and social impacts (NRC 2003). Northern Alaskan communities have become dependent on oil and gas revenues to maintain new infrastructure, modern equipment and contemporary lifestyles (Mikkelsen at al. 2008). In his environmental history of the Trans-Alaska pipeline, Peter Coates (1991) has written about the controversies surrounding the development of North Slope oil and the planning and construction of the pipeline and has argued that the politics of nature in the American West (including Alaska) have been dominated by energy issues.

With production from Prudhoe Bay having peaked some years ago, and demand for energy in the United States increasing, the search is continuing for viable alternatives to the oil produced from these vast reserves. Most Alaskan oil continues to be produced in the Prudhoe Bay area, with the Endicott, Northstar and Point

Macintyre fields being three of the largest but, since 2001, Alaska has seen a new surge in exploration for oil and gas in previously under-explored areas of the state, including several parts of the interior and the Alaska Peninsula, as well as leasing plans for the Beaufort Sea, Chukchi Sea, Norton Basin and Hope Basin (Alaska Department of Natural Resources 2006). As Mikkelsen et al. (ibid.: 140) observe, the activities of oil and gas companies in Alaska "are expanding *in* the Arctic, not *to* the Arctic".

Proposals to develop and exploit oil reserves on the northern Coastal Plain area of the Arctic National Wildlife Refuge (ANWR) continue to fuel ongoing controversy. ANWR is an ecologically sensitive area of the North Slope, often called "America's Serengeti" by environmental and conservation groups because of its abundant wildlife, which includes large mammals such as caribou, grizzly bears, wolves and polar bears. Originally established in 1960 as the 8.9-million-acre Arctic National Wildlife Range, the present size and status of ANWR was established by the U.S. Congress in 1980 in the Alaska National Interest Lands Conservation Act (ANILCA), and now includes some 19 million acres. Waterman (2005: xiii-ix) has written that there is no other place in North America with "such a diverse concentration of wildlife or such an unlikely combination of pastoral and stormbound beauty".

Bordered on the north by the Beaufort Sea, the only community within the boundaries of ANWR is the Inupiat village of Kaktovik (with a population of around 220) on Barter Island. The Gwich'in community of Arctic Village (inhabited by some 250 people) nudges the southern boundary of the Range. ANWR's lands are a critical habitat for the migratory Porcupine caribou herd, which constitutes a principal form of subsistence for both Inupiat and Gwich'in peoples. ANILCA set aside lands for national parks and wildlife refuges while making specific provisions for resource use within their boundaries, with preference given specifically to subsistence—i.e. what is usually meant to refer to customary and traditional—use of wild resources by rural residents for personal and family consumption. However, a clause in ANILCA called for "…an analysis of the impacts of oil and gas exploration, development, and production, and to authorize exploratory activity within the coastal plain in a manner that avoids significant adverse effects on the fish and wildlife and other resources" (Maas 2005: 33). A study was completed by the U.S. Fish and Wildlife Society in April 1987 and concluded that oil and gas development should be allowed and that the wildlife population on the Coastal Plain would not suffer major disruption or harm.

The Arctic Oil and Gas Association (AOGA) produced an advocacy paper in 1986 providing briefing material as to why the U.S. Congress should allow oil and gas exploration in the Coastal Plain area. It argued that the Coastal Plain "has the highest potential of any unexplored region in the onshore United States. The United States must continue to explore for and develop its petroleum resource potential in the face of increased dependence on foreign sources" (AOGA 1986: 1). It is notable that in government documents of the time, concern over American dependence on oil from the Middle East was given as one of the pressing reasons why exploration should be permitted to take place in ANWR. Reducing dependency upon imported oil by developing domestic resources, it was argued, would not only have a positive impact on the nation's trade deficit, it would ensure reliable sources of domestic oil. Opening up ANWR would not only be in the national interest, it would have significant foreign policy implications. To date, ANWR and the need to develop domestic oil and gas supplies continues to play a significant role in U.S. political debates on how the prevention and disruption of oil supplies is a key factor in U.S. foreign policy decision-making. The pro-developers have found it effective to play on fears of American vulnerability to serious economic and security crises if the country is allowed to fall into a position of complete dependency on foreign oil.

In 1989, Commonwealth North, which describes itself as a non-partisan group acting as a public policy forum in Alaska and supported by private individuals and public sector organizations, produced a report called *An Alaskan View of ANWR*. It reinforced the AOGA position, as well as the views of other groups and individuals in favour of developing ANWR, and also cited United States Geological Survey (USGS) estimates of the time that there was potential for 4.9 billion barrels of oil in the Coastal

Plain. Commonwealth North argued that the oil potential of the Coastal Plain "goes beyond mild speculation" and that it "contains the most promising onshore frontier area for major oil and gas prospects in the entire nation" (1989: 5). The AOGA argued that oil exploration was entirely consistent with ANWR management and conservation goals and that "subsistence species (primarily caribou and waterfowl) in the Coastal Plain will not be diminished by petroleum exploration and development; access to these resources would not be significantly affected by oil and gas activities" (AOGA ibid.: 7). The pro-developers argued that fears over environmental damage were unfounded. They pointed to the co-existence of caribou herds with oil and gas development and activities elsewhere in northern Alaska, arguing that the herds had actually increased their numbers in and around oil fields, and that oil development in the Kenai National Wildlife Refuge in south central Alaska had continued since 1962 without any adverse impacts to key species, such as moose or salmon (AOGA ibid.: 6).

However, environmental and conservation groups, both within Alaska and elsewhere in the United States countered pro-development arguments and continue to argue that ANWR should remain closed to exploration and development. In particular, groups such as the Fairbanks-based North Alaska Environmental Center argued that scientific research showed that oil development harms caribou and that the ecological dynamic is more complex than groups such as the AOGA were suggesting. Caribou, it was argued, do not co-exist harmoniously with the oil and gas industry. The status of caribou herds in regions of northern Alaska affected by oil development remains a contentious debate. While environmentalists, conservation groups and some scientists maintain that oil exploration and development presents an environmental risk to the Coastal Plain of ANWR—and indeed to the entire Refuge—it is also an issue that divides Alaska Native communities: for the Inupiat, oil development presents economic opportunity while for the Gwich'in (and neighbouring Gwich'in communities in Canada's Yukon Territory) it threatens cultural survival.

ANWR and Oil

Situated in north-east Alaska and managed by the U.S. Fish and Wildlife Service, the Arctic National Wildlife Refuge has only one inholding, comprising a large surface estate owned by the Inupiat Eskimo village of Kaktovik and a subsurface estate owned by the Inupiat-controlled Arctic Slope Regional Corporation (ASRC). ANWR is also one of the last regions of the U.S. Arctic (and the Coastal Plain is the only region of the North Slope) not open to oil and gas development. To the west and north of ANWR, the Alaskan state government and U.S. federal government are pursuing leasing programmes in the National Petroleum Reserve-Alaska (NPRA) and in the Alaskan part of the Beaufort Sea. A major report deriving from a study carried out by the Committee on Cumulative Environmental Effects of Oil and Gas Activities on Alaska's North Slope (NRC 2003) showed that more than 1,000 square kilometres of the North Slope have been transformed into a sprawling industrial complex. Ongoing leasing activities and advances in oil recovery technologies on the North Slope and in the Beaufort Sea mean a substantial increase in the area of northern Alaska that may be opened up for exploration and development.

Long regarded as a potential source of significant oil and gas reserves, recent U.S. Geological Survey estimates are that that the Refuge contains 10.36 billion barrels of oil, with 4.5 billion barrels under the Coastal Plain, and 12 trillion cubic feet of gas. While these are not enormous reserves when compared globally, the Refuge nonetheless has tremendous symbolic value for all sides in the long-running debate over its future. Proponents of opening up ANWR to development argue that it could represent one of the last great oil discoveries in the United States, and its development would ease U.S. dependence on oil imported from the Middle East.

Environmentalists argue that ANWR contains some of the last great wilderness areas in the country, which would be destroyed if development went ahead. The USGS has pointed out that, since the oil may not be found in one specific location but scattered in pockets across the Coastal Plain, exploration and development will require an extensive networks of roads to connect the facilities and infrastructure. Waterman (2005: xiv) has described the landscape as "continually spilling off into a limitless horizon, regardless of where you stand" and how "to know this place is to desire its protection". The Gwich'in people of Alaska and western Canada, whose subsistence lifestyle depends on the nearly 130,000-strong Porcupine caribou herd that relies on the Coastal Plain as its annual calving ground, call it "the sacred place where life begins" and are generally opposed to development. While Inupiat Eskimo hunters in northern Alaska also rely to some extent on the Porcupine caribou herd, many Inupiat—especially those in control of corporations and businesses—support exploration and would stand to benefit financially from the development by leasing lands they hold in ANWR to oil companies.

Opening up ANWR

For more than two decades, industry has lobbied for access to oil resources within the Refuge, while environmentalists continue to campaign for ANWR to remain a wilderness with no development within its borders. As political scientists McBeath and Morehouse (1994: 1) wrote, ANWR "pits the interests of Alaska economic development against those of national environmental conservation. It also juxtaposes the consensus of opinion in the United States, which has favoured preserving the Refuge, against the will of a majority of Alaskans who look to ANWR for future economic security." Pro-developers disagree that the term "wilderness" can be used to describe ANWR's environment. They argue that the area cannot be represented as untouched and pristine. People have lived there for millennia, the land has been used by indigenous hunters, fishers and trappers, and resources have been exploited by more recent historic incursions into the area by commercial whalers, trappers and traders. For instance, from around 1890 to 1910, Barter Island was a key node in a network of trading places during the era of commercial whaling. In the 1920s, trading posts for fox furs were set up and several reindeer herding ventures were tried in the ANWR area. The H.B. Liebes Company of San Francisco established a trading post at Barter Island in 1923, but the market for fox furs collapsed in the late 1930s and, together with the starvation of many reindeer herds during severe winters around the same time, the residents of the Kaktovik area suffered economic hardship. In 1936 many were reported by the Bureau of Indian Affairs to be facing starvation. Some of the local residents were able to obtain wage employment during U.S. Coast and Geodetic Survey mapping projects of the Beaufort Sea coast in the mid-1940s, and Kaktovik became the site of a U.S. Air Force runway and hangar in 1947, as well as the site of a Distant Early Warning (DEW) line facility a few years later. Against this background of historic use and human activity, and with careful planning, pro-developers argue that there is no reason to suppose that oil and gas development will diminish the aesthetic value of ANWR and the Coastal Plain (AOGA ibid.)

The U.S. Congress has long attempted to balance its desire to preserve ANWR as an ecologically-rich area with the need to explore its potential as an oil-rich frontier. This is reflected in section 1002 of ANILCA, in which Congress requested a report and recommendation on development. ANWR's Coastal Plain (called the 1002 Area) consists of 1.5 million acres (which is approximately 10% of ANWR's total acreage). The "1002 report" (and hence the Coastal Plain's 1002 Area appellation), as it became known, was submitted to Congress on 1 June 1987, and it recommended for the first time that Congress should enact legislation to open the Coastal Plain up for oil and gas exploration and development. A later U.S. Fish and Wildlife Service report argues, however, that the 1002 Area is critically important for the ecological integrity of the entire Refuge (U.S. Fish and Wildlife Service 2001).

Since then, attempts have been made to push a number of bills through U.S. Congress, some aiming to implement the recommendation, others intending to ban development and declare the Coastal Plain a wilderness area. Until recently, moves to have ANWR opened to industry had been unsuccessful within the U.S. government. Motions were defeated by the Democrats under the Clinton administration but received greater support under George

W. Bush's two terms in office. In May 2005, the U.S. Congress voted in favour of allowing drilling within the refuge, by way of approving a Budget Resolution containing a provision to open the Refuge up through the annual budget process rather than through energy policy legislation. In October 2005, the Senate Energy Committee voted to open ANWR to oil drilling as part of a broad budget plan, yet two months later the U.S. Senate voted against drilling. For now, the Arctic National Wildlife Refuge remains the only area on Alaska's North Slope where oil and gas development is specifically prohibited. The rest of the North Slope is available for oil and gas development through decisions made by the Secretary of State of the Interior for the National Petroleum Reserve-Alaska and the Beaufort Sea, and by the Commissioner of the Alaska Department of Natural Resources for state lands and waters (U.S. Fish and Wildlife Service ibid.). That said, by the mid-1980s, the Department of the Interior had already divided ANWR's Coastal Plain into tracts of some 2,500 acres each and this parcelization of the area in anticipation of future exploration is similar to that used for federal oil and gas lease sales.

Alaska Native Interests in Oil and Gas

The discovery of oil at Prudhoe Bay on Alaska's North Slope in 1968 was the latest in a series of development projects which had been experienced by indigenous people through the decade as an encroachment on traditional lands and it led to demands for land claims by the Alaska Federation of Natives (AFN). In 1971 the United States Congress passed the Alaska Native Claims Settlement Act (ANCSA) which, while not recognizing a Native land claim to the whole of Alaska, nonetheless established 12 regional Native corporations effectively giving them control over one-ninth of the state (one further corporation was established for non-resident Alaska Natives). It was the last Native land claim settlement to be reached in the continental United States and, at the time, was described as the first modern treaty in North America, providing a model as well as an inspiration for future indigenous land claims settlements, especially in Canada (Colt and Pretes 2005). ANCSA transferred 44 million acres of land and US$962.5 million to business corporations owned exclusively by Alaska Natives. Today, many of these corporations are involved in some way in the oil and gas industry. For example, the Ahtna Construction and Primary Company is involved in oil spill response and pipeline work; the Arctic Slope Regional Corporation (ASRC), which is based in Barrow, includes Alaska Native-run oil and gas companies; and Doyon Ltd. and Cook Inlet Region Inc. both provide various oil field support services. In terms of training, in addition to the opportunities provided by Native corporations, the Anchorage-based First Alaskans Institute offers summer internship positions that can be related to opportunities in the oil and gas sector.

The oil and gas fields of northern Alaska play a major role for Alaska Natives. For example, the Northwest Alaska Native Association (NANA) is one corporation doing well out of oil field services, catering services, hotels, power generation and distribution to the major oil and gas companies in Alaska. NANA also has a joint venture with the Comino mining company for the extraction of zinc and lead from their lands. Doyon Limited, which represents the Alaska Natives of the interior, operates Doyon Drilling and Doyon Universal Services JV in support of oil and gas development throughout Alaska. Ongoing leasing activities and advancement of oil recovery technologies on Alaska's North Slope continue to provide new opportunities for exploration and development support, areas in which Alaska Native corporations are key players.

As I argued in Chapter Three, comprehensive land claims agreements have made Aboriginal business ventures in the Northwest Territories closer to their counterparts in Alaska than elsewhere in northern Canada. Northern Canadian Aboriginal-owned corporations resemble Alaskan Native corporations in both their institutional culture and business ambitions. Officials of the Arctic Slope Regional Corporation (ASRC) have argued that the discovery, extraction and transportation of oil from Alaska's North Slope provides an excellent example of the energy industry and indigenous people developing an understanding, engaging in dialogue, working together on industry and indigenous issues and finding solutions to the benefit of industry and Alaska's Native people. ANCSA paved the way for the Trans-Alaska Pipeline System but did not automatically result in contracts and jobs for Alaska Natives. The contract and performance situation is very different today—for example, ASRC companies now take on contract work in oil production and pipeline maintenance worth several hundred million US dollars. ASRC has petroleum contractor subsidiaries but has moved into refining and distribution, with oil refineries in Valdez in southern Alaska and the town of North Pole, near Fairbanks, as well as gas stations and diesel distributions all over Alaska. ANILCA allows for land exchanges under certain conditions and ASRC has also exchanged some of its land for land within the National Petroleum Reserve. With plans for development of North Slope natural gas and how to take it to market, ASRC's oil and gas subsidiary Natchiq is involved with teams responsible for the design of the gas processing facility and the construction of the proposed pipeline. As Richard Glenn, ASRC's Vice-President (Lands) has put it:

Our corporation, with established subsidiaries in oilfield construction, surveying and engineering, and pipeline operations, has much to contribute to the construction and operation of a natural gas pipeline. We are already contributing, for example in the "front-end engineering and design" for a portion of the pipeline along its proposed route through Canada. We seek continued participation in the design, construction, and future operations of this major development project. Our companies are competent, they have proven themselves in industry, and most importantly they seek to put our people to work.[47]

While the Arctic Slope Regional Corporation supports the opening up of ANWR, and particularly the coastal plain, to industry, it does not necessarily mean that Inupiat people overwhelmingly favour development. They, along with residents of other Alaskan communities, look to energy development as a source of jobs, the construction and running of schools, and other opportunities and, while some may be opposed to drilling in ANWR, it nonetheless remains an issue on which there is a diversity of local opinion. In anticipation of a hoped-for decision by Congress to open ANWR's Coastal Plain to oil and gas exploration and development in the 1980s, the U.S. Department of the Interior (DOI) engaged in land exchange negotiations with several Native corporations. The U.S. Fish and Wildlife Service proposed trading oil and gas rights within the coastal plain for ANCSA village and regional corporation lands within or adjacent to parts of the national wildlife refuge system in Alaska. Land exchange negotiations between the DOI and several ANCSA corporations resulted in an agreement that would allow them to make oil and gas tract selections in the coastal plain. These changes, which were written into ANILCA, allowed ASRC to swap some of its land for land not only in the National Petroleum Reserve but also in ANWR.

Alaska's North Slope Borough is often cited as a positive example of what can happen when Arctic residents have opportunities to capture some of the economic benefits from industrial development, both through employment and corporate investments. Benefits in the form of improved public infrastructure, educational services and health care can be significant. Trade-offs can be decreased where communities of resource users are afforded a significant degree of authority over development planning and operation policies to ensure that community concerns are adequately addressed. Yet oil development also brings its own dilemma of how best to balance the economic benefits with the major social changes and cultural impacts such development brings. Despite the technological transformation of the northern Alaska environment and the dominance of the oil and gas industry, cultural traditions remain strong on the North Slope and local environmental concerns are expressed when new development plans are unveiled. The Inupiat have a nutritional, cultural and spiritual relationship with the bowhead whale and other marine mammals that are threatened by current and projected oil and gas activity. Research has shown how noise from exploratory drilling and seismic exploration activities in the Beaufort Sea, for example, has disturbed bowhead whale migration routes. This forces hunters to travel further in search of their prey and exposes them to greater risks (NRC 2003). Many

North Slope residents, especially the people of Kaktovik, have expressed opposition to offshore exploration (Mikkelsen et al. ibid.) while they may have supported drilling on the Coastal Plain of ANWR.

Caribou People

The Porcupine caribou herd is the eight largest migratory caribou herd in North America. It spends each winter in northern Canada, in the Northwest Territories' Richardson Mountains and in central Yukon, and in north-eastern Alaska. The herd moves west and north during spring to its calving grounds on ANWR's Coastal Plain. Biologists tend to believe that the caribou make the journey to give birth on the coastal plain because it is safer habitat—there are fewer predators, and rich tundra plants provide a critical source of nourishment for calves and nursing caribou cows. In late June and July, the herd disperses in groups of tens of thousands of animals and continues its annual migration south and east between Canada and Alaska during autumn and winter. The management of the herd is overseen by the Porcupine Caribou Management Board and, because it moves across the Canada-U.S. border during its migration, the board was mandated under an international agreement between the two countries that has been in force since 1987. The agreement was signed by the federal governments of the United States and Canada, the territorial governments of the Northwest Territories and Yukon, and organizations representing Yukon First Nations, Inuvialuit and Gwich'in.

Gwich'in have relied on the Porcupine caribou herd to meet essential subsistence, nutritional, cultural and spiritual needs for thousands of years. The Gwich'in Steering Committee was established at a meeting in Arctic Village in response to the possibility of the 1002 Area being opened up to development, and it acts to represent the rights and interests of the Gwich'in people who currently make their home on or near the migratory route of the Porcupine caribou herd in communities in Alaska, Yukon and the Northwest Territories.[48] The Committee asserts that opening ANWR up constitutes a threat to the caribou calving grounds, which in turn is a threat to the very heart of the Gwich'in as a people. The Gwich'in Steering Committee was established with a resolution, Gwich'in Niintsyaa, proclaiming the inherent right to their means of subsistence, and asserting that oil development brings the real threat of endangering Gwich'in society and culture.[49]

As with the Inupiat concerns about offshore development affecting bowhead whales, the Gwich'in worry about oil development disturbing herd reproductive and migratory behaviour. These concerns are intense and widespread in Gwich'in communities and are backed up by research, which has already shown that caribou are sensitive to disturbance during calving (Griffith et al. 2002). It is clear that oil development in the 1002 Area would potentially impact on the Porcupine caribou herd. Infrastructure development, in terms of pipelines, seismic trails, access roads, well-pads and other structures, is likely to reduce the amount and quality of forage for caribou during and after calving, restrict access to insect-relief habitats, expose the herd to higher predation and affect the herd's migratory pattern (U.S. Fish and Wildlife Service 2001).

Canada opposes oil development in ANWR. An agreement signed in 1987 between Canada and the United States recognizes that the two countries have a joint responsibility to oversee the habitat of the herd and to protect the calving grounds. Indeed, ANWR is a critically important part of a larger international network of protected Arctic and sub-Arctic areas. In the northern part of Canada's Yukon Territory, the Canadian federal government worked with First Nations to establish Ivvavik and Vuntut National Parks, two areas that border ANWR and in which oil and gas exploration and production are banned.

As the only group that lives within the boundaries of the Refuge, the Inupiat residents of Kaktovik claim that it should be their opinion that takes precedence over groups living outside of the Refuge. ANWR here becomes an issue of stewardship, with the Inupiat arguing that they are knowledgeable enough about the land to make decisions regarding development. They advance the claims that North Slope development did not have the devastating effect on wildlife that was anticipated, and the Inupiat now know that industry can be responsible and coexist well with the environment. The Inupiat assert that they, too, rely on the Porcupine caribou herd, would not wish to see it threatened and argue that they have seen elsewhere on the North Slope that caribou and industry can coexist successfully.

Are there real benefits to developing the 1002 Area?

The debate over ANWR is not just about wilderness preservation and cultural survival in a small corner of the Arctic. ANWR's potential as a major source of energy is advanced by advocates of development as a national security issue, whereby oil from ANWR will be crucial to ease U.S. dependence on foreign oil at a time of increasing U.S. oil consumption. Opponents of ANWR development, including the Gwich'in Steering Committee, argue that opening the 1002 Area to drilling would be a fiscally irresponsible decision, since there is no way of yet knowing—despite all estimates—how much oil is available. Environmental groups also argue that the financial costs incurred in exploration and development may not be recovered from the oil reserves, which may not be as significant as hoped. In the 1990s, for example, it was claimed that a U.S. Geological Survey report, which gave a low figure for reserves in ANWR, was withdrawn under pressure from Alaskan politicians and rewritten with a slightly more optimistic conclusion (Roberts 2004). And as Ricki Ott points out in a recent book on the *Exxon Valdez* disaster in Prince William Sound in 1989, when the tanker hit a reef spilling 11 million gallons of crude oil into the waters (Ott 2005), what then were the environmental and human health consequences of U.S. dependence on oil? While Ott discusses the environmental impact of the *Exxon Valdez* accident, she also shows how workers involved in the clean-up operation developed chronic illnesses. She also claims that studies show that between 50-100 tons of oil remain in Prince William Sound and that oil continues to harm wildlife and ecosystems for at least 15 years after a spill has happened.

Between 1996 and 2004, exploration and production operations in the sprawling Prudhoe Bay complex resulted in an average of more than 500 reported oil spills annually.[50] In March 2006, around 6,400 barrels of oil leaked from a corroded transit pipe at BP Alaska's operation at Prudhoe Bay, forcing the company to temporarily shut down production of 400,000 barrels a day. Such incidents, as well as the continuing legacy of the *Exxon Valdez*, as well as other oil spills and leaks elsewhere in the United States such as in the Gulf of Mexico, continue to focus attention and concern on Alaska's dependence on oil and act as a constant reminder that what has already happened in the North could very well happen again. For the Gwich'in, environmentalists and others campaigning to keep industry out of ANWR, such uncertainty cannot justify damage to the land, its wildlife and ecosystem integrity, nor to the culture of peoples dependent on the Porcupine caribou herd.

CHAPTER VI

WATER, THE THICK BLACK OIL AND THE GATEWAY TO ASIA

The Peace-Athabasca Delta in north-eastern Alberta's sub-Arctic boreal region is the world's largest inland freshwater delta. It is a landscape on a grand scale, the meeting place of two of Canada's mightiest north-flowing rivers, the Peace and the Athabasca, which wash in nutrient-rich sediment over the delta in an annual flood cycle. From there, the waters of these two rivers enter Lake Athabasca, then continue into Great Slave Lake in the Northwest Territories via the Slave River. From Great Slave Lake, the waters of the Peace and Athabasca meet the Mackenzie River and eventually flow into the Arctic Ocean. The Peace River rises in northern British Columbia, while the Athabasca River originates from the Athabasca Glacier, one of the eight principal glaciers fed by the Columbia Icefield, which straddles the Continental Divide in the Rocky Mountains. The Columbia Icefield is the hydrographic centre of Canada. It is about 325 km² in area, 100 to 365 metres in depth, and receives up to seven metres of snowfall per year. The Columbia Icefield and northern Alberta's rivers have assumed environmental, scientific and cultural significance for debates about water security and community well-being and intersect with discussion of the environmental, social and health impacts of energy development. As two of the main tributaries of the Mackenzie River, the Peace and Athabasca rivers play a major role in defining the water resource-related constraints in north-eastern British Columbia and northern Alberta. They are a source of life and livelihood for the people who live along or near their banks. They are also central to the story of the exploration and development of northern oil and gas reserves, visions of the energy future, and how all this affects indigenous and local communities today in an enormous area of north-western North America.

An Ecosystem in Crisis?

The Peace-Athabasca Delta is a vast wetland that attracts thousands of nesting birds, including the threatened whooping crane, and is critically-important habitat for many other species of wildlife, including beaver and muskrat and a diversity of fish species. It was designated a World Heritage Site in 1985. The convergence of the Peace and Athabasca river systems in this ecologically-sensitive area has provided Aboriginal peoples with the means to sustain livelihoods, societies and cultures based on fishing, hunting, and trapping for around 7,000 years, but the Peace and the Athabasca are also vital in the way they have offered other indigenous peoples and, later, European settlers the means to create viable communities in the Mackenzie Basin more widely. Fort Chipewyan is the closest community to the delta and was established in 1788 as the first European settlement in Alberta. Situated at the hub of the Mackenzie Basin drainage system, it was ideally situated for the fur trade and became a base of operations for land explorers, such as Alexander Mackenzie, John Franklin, George Back and John Richardson, among others. The viability of the community of Fort Chipewyan and the ecological integrity of the delta are threatened today by the effects of the Bennett Dam on the Peace River as well as by the development of the Athabasca oilsands a little further to the south.

Over the last 30 years, the Fort Chipewyan First Nation, scientists and environmentalists have been observing that the Peace-Athabasca Delta has been drying up and Aboriginal people have described how the changes are affecting the region and local livelihoods (e.g. Campbell 2000). The reduction in annual flood levels is cited as part of the explanation as to why these ecologically-critical wetlands have been drying, but the reason for this reduction is also partly the result of the way in which water has been managed since 1968 as a result of the construction of the W.A.C. Bennett Hydroelectric Dam in the Rocky Mountains on the British Columbia stretch of the Peace River.

Built by B.C. Hydro, the W.A.C. Bennett Dam was a controversial project with significant environmental and socio-economic impacts. Initiated by and named after W.A.C. Bennett, who served as Premier of British Columbia from 1952 to 1972, the dam was the realization of dreams, held by many politicians and industry leaders since the early 20th century, about the energy potential of the northern British Columbia and northern Alberta frontiers. Construction began in 1962 and was completed in 1967 and the dam created the reservoir of Williston Lake, which took five years to fill. The dam and its construction caused considerable displacement of both indigenous people and settlers of European descent, yet no assessments were carried out prior to construction to consider the potential impacts of the dam project on people and their communities and livelihoods (Brody 2000).

Howell (1978) described the political and economic decision-making processes which led to construction of the dam and argued that there was a notable absence of balanced decision-making at critical stages of the project's development. In the face of local concern about its impacts, he discussed various legal approaches to the provision of compensation to communities adversely affected by construction of the dam. Brody (ibid.) has related how families with farms in the reservoir area were given compensation based only on current land valuations of the time. The long-term impacts on the social and economic lives of those displaced were not considered. The large areas which were flooded were also economically and culturally significant to trappers (most but not all of whom were Aboriginal) and this received some attention. This was a region where traplines—a term used to describe areas which are licensed to individuals for the trapping of fur-bearing animals for marketing purposes—had been registered in the 1920s and 1930s. In 1965, as the reservoir began to fill, people were relocated. Pollon and Matheson (1989) chronicled how the indigenous Sekanni, who lived at the settlement of Ingenika and hunted, trapped and fished in the valley that was eventually flooded, were not consulted beforehand. People with traplines were merely told that the land would be flooded and that they would receive compensation. Initial payments of Can$2,700 were made to each family having a trapline affected by the development and subsequent flooding. Of this money, each family received $100 to $200 cash, with the remainder "administered" by the appointed (non-native) Indian Agent (Brody ibid.).

Oral testimony reveals that the Sekanni were told their houses would be burned but that their belongings would be put in a safe place:

Albert Poole's father was a trapper. Half of his trapline is under-water now. He was one of the people who found his house burned with everything in it, Poole said. "He came home one day to find his cabin burned. I was in Finlay Forks at the time it happened. Everything was burned: guns, pictures, all that. It was all gone." Poole said his father received $2,700 compensation for his flooded traplines, and that's all.

Pollon and Matheson (1989: 338)

Eventually, in 1987 the Sekanni people living at Ingenika at the time of the flooding were awarded a total of Can$180,000 by the provincial government. Hugh Brody has written that, when the Bennett Dam project went ahead, it was probably the case that very little was known about traditional land use in the area by indigenous people:

The dam was built, and flooding took place, before any land use and occupancy studies—the body of research that in many parts of Canada has created baseline data for impact assessment for indigenous peoples— had been carried out. Also, the dam was built in an era of Canadian administration of "Indian Affairs" when large-scale development was rarely if ever challenged or even modified by indigenous interests. This was part of a widespread faith in such projects, and a profound confidence in their social value—both locally and, more decisively, to the province or nation as a whole. The macro economic and macro social interests were interlinked, and the project welcomed as more or less unquestionably good.

(Brody ibid.: 5)

Soon after its construction, the effects of the dam on Lake Athabasca, particularly the reduced water levels of the delta, were being noted by scientists and local people. A series of hydrological studies were carried out that argued the necessity for introducing measures to restore the level of the lake so that it would fluctuate within an acceptable range and achieve the appropriate timing of peak levels (e.g. Coulson 1969, Card and Yaremko 1970).

In 1987, the Peace-Athabasca Delta Implementation Committee, an intergovernmental committee representing the governments of Canada, Alberta and Saskatchewan, produced a report which related to discussion about restoring the low water levels that had occurred on the delta following construction of the Bennett Dam. The Northern River Basins Study (NRBS) of 1996 drew on scientific findings that show that the Bennett Dam has regulated flow in the Peace River since 1968, altering the physical characteristics of the river and influencing ice formation and break-up as far downstream as the Peace-Athabasca Delta. The Peace River is normally ice-affected for about seven months of the year, in one reach or another, between the dam and the Slave River. The river's ice regime, as well as seasonal freeze-up, is influenced by both the climate and the discharge into the river, with the latter depending on the way Williston Lake is managed. Changes in either climate or the discharge into the river will have dramatic effects on the timing of freeze-up and, ultimately, on the Peace River's ice regime for the entire year.

Beltaos (2003) showed that ice jamming during the spring breakup of the ice cover in the lower reaches of the Peace River is the main agent of flooding and replenishment of the Peace–Athabasca Delta ecosystem. The relative rarity of major ice jams in the lower Peace River following construction of the Bennett Dam has resulted in serious habitat degradation and risk to local ecology, and concern has been raised over potential climate change impacts. The Intergovernmental Panel on Climate Change's (IPCC) 3rd and 4th global assessments show how Arctic and sub-Arctic river systems are particularly susceptible to current and predicted climate change. For the Peace and Athabasca rivers, a warmer climate will likely create a more pluvial runoff regime as a greater proportion of the annual precipitation will come from rain rather than snow (Anisimov and Fitzharris et al. 2001). According to the science, climate warming will lead to a shortened ice season and thinner ice cover, yet will result in a situation where both the Peace River and the Athabasca River will be prone to ice jamming and hence larger annual flood peaks. The micro-climate will change in some areas of the Peace and Athabasca river valley because of the possibility that stable ice cover may not form.

The Bennett Dam, research shows, dampens the high and low flows of the river to ensure peak energy generation potential during the winter months when market demand for electricity is highest. Moreover, the water that passes through the dam is drawn from the lower portion of Williston Lake, an area remaining unfrozen during the winter months. These changes in water levels and temperature can alter many aspects of the ecosystem, such as the quantity of habitat, the movements of fish and animals, and the period in which the river remains frozen. Discussing results from the NRBS, Prowse et al. (2002) show that this regulation of the Peace River has shifted the pattern of seasonal flows and damped flow extremes, creating a situation where the water and ice show less variability annually. Furthermore, they argue that increased winter releases from Williston Lake have virtually eliminated the formation of a complete winter ice cover for a significant distance below the dam and have delayed the formation of river ice further downstream.

The river regulation effects of the dam have been shown to have negative impacts on the economy of Fort Chipewyan, which is some 1,200 kilometres downriver of the dam. It has been estimated that direct losses from traditional economic activities such as trapping, hunting, fishing and gathering for local food production and consumption have ranged from around Can$112,000 to $210,000 per year (Adams 1998, Brody ibid.). Environmentalists and concerned residents argue that the Bennett Dam remains a threat to the Peace-Athabasca Delta. Wrona et al. (2000) have shown that the northern river ecosystems of the Peace, Athabasca and Slave are under increasing environmental stress from development activities that influence water quality and associated ecological integrity. In particular, contaminant-related threats to river ecosystems arise from increased industrial activity as well as land-use activities such as forestry, agriculture and mining.

Timoney (2002: 297) argues that, despite increasing concern that it is an ecosystem under threat, the watershed is nonetheless not as profoundly disturbed as most other major river systems in North America. He points out that the impact of the Bennett Dam must be seen in a broader context of development. The forest and petroleum industries, agriculture and changing climate, he says, are more pervasive in their ecological impacts on the delta, and the effects of accumulating contaminants and increased industrial development further upriver may constitute the major threats. In a later report, Timoney (2008) concluded that the people and biota of the Peace-Athabasca Delta and the western part of Lake Athabasca were exposed to higher levels of some contaminants than those upstream. Concentrations of arsenic, mercury and polycyclic aromatic hydrocarbons (PAHs) are already high and appear to be rising. Because they consume traditional foods produced from hunting and fishing activities, First Nations of the area are exposed to high risks. Many of these risks originate from pollutants produced from the oilsands mines north of Fort McMurray. The production process in the region, turning bitumen into synthetic crude oil, has an enormous environmental impact on the Mackenzie Basin and threatens both the water quality and water quantity of the Athabasca River and other rivers and freshwater lakes.

Oilsands Development, Arctic Gas and the Alberta Hub

In 1873, George M. Grant, chronicler of the expedition across Canada led by Sandford Fleming the previous year, was enthusiastic in his assessment of the potential of the Northwest for resource development and settlement:

> ...it is impossible to avoid the conclusion that we have a great and fertile North-west, a thousand miles long and from one to four hundred miles broad, capable of containing a population of millions. It is a fair land; rich in furs and fish, in treasures of the

forest, the field, and the mine; seamed by navigable rivers, inter-
laced by numerous creeks, and beautified with a thousand lakes;
broken by swelling uplands, wooded hill-sides and bold ridges; and
protected on its exposed sides by a great desert or by giant moun-
tains....Here we have a home for our own surplus population and
for the stream of emigration that runs from northern and central
Europe to America. Let it be opened up to the world by rail and
steamboat.

(Grant 1873: 179-80)

Thirty-five years later in *Through the Mackenzie Basin*, Charles Mair wrote of his travels in the Athabasca River region with the Treaty 8 commission in 1899 as taking him to "perhaps the most interesting region in all of the North". He was astonished by the "impressive grandeur" of the "tar-cliffs" and the "tar-wells" and described the tar as "a fuel which burned in our campfires like coal" (Mair 1908: 121). He was writing about the bitumen first recorded by Peter Pond and Alexander Mackenzie in the 18th century and used by indigenous peoples of the region for gumming canoes and boats. Mair reinforced the view prevalent at the time of his journey that this viscous mix of oil, silica sand, clay, minerals and water was of significance to Canada's future:

That this region is stored with a substance of great economic value
is beyond all doubt, and, when the hour of development comes, it
will, I believe, prove to be one of the wonders of Northern Canada.
We were all deeply impressed by this scene of Nature's chemistry,
and realized what a vast storehouse of not only hidden but exposed
resources we possess in this enormous country.

(Mair ibid.)

Mair's view was endorsed by many other writers in the early years of the 20th century. R.G. MacBeth, author of works on opening up the West, such as *The Making of the Canadian West*, and *Selkirk Settlers in Real Life*, marvelled at the expansion of agriculture in the Peace River district and the resource potential of northern Alberta and British Columbia. In *The Romance of Western Canada*, he observed that "...we see how the map of Canada has been rolling backward until, by degrees, we have come to understand that this Dominion is possessed of a country so vast in extent and so rich in resources that we have hardly begun to understand the illimitable material possibilities that lie within our borders" (MacBeth 1918: 236).

The Athabasca region contains one of three deposits of oilsands in Alberta—the other two being located near Peace River and Cold Lake. Suncor and Syncrude are the two main operators, although other oil companies such as Shell, Petro-Canada, ConocoPhillips, Husky and Imperial have interests in oilsands projects. The development of the technology to extract the bitumen and refine it into crude oil meant that production in the Athabasca oilsands did not begin until the 1960s. Today, oilsands mining activities in northern Alberta continue to expand and the oilsands are estimated to lie under a total of 4.3 million hectares, or some 10.6 million acres, with enough in-place bitumen to produce 1.7 trillion barrels (the remaining established reserve of bitumen is 172.3 billion barrels, compared with about 1.6 billion barrels of conventional oil left in the province). This makes Canada's potential oil reserves from the oilsands alone second only to Saudi Arabia's oil reserves. Something like 1.25 million barrels of oil are produced from oilsands operations every day, while conventional oil production in Alberta delivers over 500,000 barrels per day.

With relatively accessible conventional oil supplies beginning to run dry, governments and energy companies are increasingly attracted to Alberta's oilsands. The Alberta Energy and Utilities Board (AEUB) *Supply/Demand Outlook 2007-2016* report says that Alberta's production of bitumen could triple by 2016 and will account for 75% of the province's total oil production. By that time, the report forecasts, Alberta will have become one of the world's largest oil producers. Much of that oil will probably go to the United States—the U.S. Department of Energy has predicted that crude oil from Alberta's oilsands will help halve America's dependence on imported oil from outside North America within two decades (Alberts 2006), yet India and China also constitute future markets. The intensifying activity in the oilsands is resulting in an expansion of pipeline projects, with two major oil-export pipelines due to go into operation later in 2010. One of the major pipelines for bitumen is TransCanada's Keystone pipeline, a Can$12 billion project which is initially moving 200,000 barrels a day from northern Alberta to Illinois.[51]

The hunger for energy experienced in North American and Asian oil and gas markets is one reason why industry is concerned to get regulatory approval to construct pipelines to move Arctic gas from Alaska and the Mackenzie Delta. A concept called the "Alberta Hub" was introduced by the provincial government of Alberta several years ago. It envisages a network of Arctic pipelines connecting to existing infrastructure in Alberta and aims to make the province attractive to Alaska natural gas producers (Tulk 2005). At the same time, while the concept constructs an image of Alberta as a conduit through which Arctic gas passes on its way to southern markets, capturing some of that gas for use in oilsands production is part of the strategy.

However, oilsands production is an extremely costly business and is hydrocarbon and water intensive. It is a strip-mining operation and the extraction process is expensive and complex, involving the mining of the resource from open-pits followed by the separation of sand, water and bitumen by using heated water and hydrocarbons. Once extracted, the bitumen is then cleaned, processed and refined, but in the process of extraction and production the industry contributes to significant environmental damage (Marsden 2007, Nikiforuk 2008) and is projected to produce 100 million tonnes of CO_2 by 2012. Currently, Alberta's oilsands industry accounts for 5% of Canada's CO_2 emissions. It takes three to five barrels of water to make one barrel of oil and this water is drawn mainly from the Athabasca River—the wastewater then ends up in toxic tailing ponds.[52] If production is to reach 3 million barrels a day by 2016 it will require almost three times as much natural gas to be used to recover the bitumen and then upgrade it to make synthetic crude. A report by the National Energy Board produced in 2006 suggested that natural gas requirements for the oilsands industry were projected to increase substantially from 0.7 billion cubic feet per day in the mid-2000s to 2.1 billion cubic feet per day in 2015 (National Energy Board 2006b). While high natural gas prices have encouraged oilsands operators to use gas more efficiently and to look for alternative fuels, the report points out, it is nonetheless the case that near future development will still require extensive supplies of natural gas.

Expansion of oilsands activities and production will mean that much of the natural gas from the Mackenzie Delta will end up in the Fort McMurray area if a pipeline is built along the Mackenzie Valley. In addition to concerns over river water, a report published in 2009 by the Council of Canadian Academies concluded that Alberta's groundwater resources had also been changed significantly by oilsands development. The extraction technology requires a nearby water resource and over 88 million barrels of freshwater a year are used by oil companies in the area. The report suggests that demand for groundwater could be greater than the demand for surface water, especially as *in situ* development of bitumen reserves becomes more common as an alternative to strip mining. This method requires considerable supplies of water to pump the bitumen to the surface.

The enormous deposits of oilsands which are being mined in a process of utter and complete environmental transformation of the boreal northlands lie underneath the traditional territories of Cree and Dene First Nations, as well as Métis communities. In the Athabasca oilsands area, some eight communities within the Treaty 8 area are directly impacted by active development of the resource (Westman 2006). Earley (2003) also found that the oilsands industry was placing considerable pressure on Fort McMurray, a city of some 73,000 inhabitants and the only urban area within 350 km of the major oilsands operations. The social impacts of the oil boom experienced in Fort McMurray—which has had an annual population increase of over 8% during the last decade—include high housing prices, shortages in service industry labour, insufficient social services, at times, to assist individuals and families who can no longer cope with the difficult conditions in the area, and a variety of other negative effects. The concerns of First Nations communities relate to environmental disturbance, air and water pollution, and habitat loss, as well as the impact of oilsands activities on human health and community well-being.

Aboriginal people do derive some benefits from oilsands development, however. For example, around 1,200 Aboriginal people find employment in the oilsands mines and related industries and Aboriginal companies land lucrative contracts with oil companies. Other community initiatives benefit from direct funding from oil companies—for example, Syncrude donated a large amount of money to create an elders facility in the Cree community of Fort McKay, and communities receive support for literacy programmes, community development projects and training programmes. The Fort McKay First Nation have also negotiated with industry for a number of potential joint venture projects although, as Westman (ibid.) points out, this does not reflect a regional consensus among Aboriginal peoples. Consultation with Aboriginal peoples about oilsands development can hardly be said to take place and environmental assessments are virtually absent. From the perspective of government and industry, it is often assumed that consultation happened at the moment that treaties were signed, thereby obviating the need for consultation prior to every new development.

The North Central Corridor Pipeline

As I discussed in Chapter Four, gas from the Mackenzie Delta would be transported to the oilsands extraction and production areas via TransCanada's North Central Corridor Pipeline. This pipeline is not only of considerable concern to the Dene Tha', it crosses the traditional territory north of Lesser Slave Lake of the Lubicon Cree, a First Nation of some 500 people. The Lubicon were overlooked by the Treaty 8 commission in 1899 because they did not live near the water routes being surveyed by the treaty commissioners. Their lands were later considered Crown lands by both the federal government and the government of Alberta and the Lubicon have been engaged in a longstanding struggle to gain recognition of their rights to land and resources—negotiations between the Lubicon and both the federal government and the government of Alberta have remained open-ended, yet the Lubicon have experienced profound social and economic change and poverty as a result of the erosion of traditional activities, logging, and oil and gas development. Commentators have written that government neglect, loss of land, and the impact of oil and gas development have contributed to a process of eliminating the Lubicon that has verged on genocide (e.g. see Churchill 2002, for a comprehensive discussion).

From a Lubicon perspective, they have retained their rights to their traditional territory because they have never signed a treaty. The downside to not having signed a treaty is that they are not recognized as being the legitimate and rightful inhabitants of the land (Espiritu 1997) and cannot evoke the spirit of treaty-signing as other groups such as the Dene Tha' and Dehcho have done to great symbolic, political and legal effect. Since the early 1980s, under the leadership of Chief Bernard Ominayak, the Lubicon Lake Indian Band have been articulating their land rights "in terms of the territory historically used by their ancestors for purposes of hunting, fishing, trapping, occupancy, and trading purposes" (Churchill ibid.: 201).

The Lubicon case has been raised repeatedly in United Nations bodies concerned with human rights. In March 1990, the UN Human Rights Committee ruled that Canada had violated the rights of the Lubicon Cree, and the right to culture as protected by the International Covenant on Civil and Political Rights, based on evidence that Canada had not recognized Lubicon land rights, especially in the face of oil and gas development. Marking 20 years of the ruling, several human rights and indigenous rights groups, environmental groups and other civic organizations, including IWGIA, Amnesty International, Cultural Survival, Tebtebba and Greenpeace Canada, issued a joint public statement on 10 March 2010 in which they argued that the Lubicon continue to suffer from the same human rights violations condemned by the Human Rights Committee. The statement expressed concern that Canada had yet to conclude a negotiated settlement with the Lubicon Cree. It called on both the federal government and the Alberta provincial government to make a public commitment to engage in negotiations on all outstanding land disputes, stating that: "No resource development activities should be permitted anywhere in the disputed land except with the clearly expressed free, prior and informed consent of the Lubicon people, as current developments in human rights law are clearly indicating."[53]

In a letter dated 15 August 2008 and addressed to Canada's UN representative in Geneva, the UN Committee on the Elimination of Racial Discrimination questioned the legitimacy of the Alberta government's right to authorize TransCanada's pipeline without local consultation and the consent of the Lubicon (Preston 2009). Notwithstanding UN concern over Canada's human rights violations in Lubicon territory, and in the absence of an agreement with the Lubicon, the Alberta Utilities Commission approved construction of the pipeline and TransCanada began clearing land for a 600-person strong contractor camp within Lubicon traditional territory in early November 2008. TransCanada has received all necessary permits and approvals required to construct and operate the pipeline despite opposition from the Lubicon Cree.

In a letter dated 13 November 2008 from Lubicon Lake Indian Nation to TransCanada, Lubicon councillors Alphonse Ominayak, Dwight Gladue and Larry Ominayak suggested a number of ways in which the company and the Lubicon Nation could engage in a comprehensive discussion about how they could work together during the construction and operation of the pipeline, and they emphasized the importance of TransCanada being able to offer the Lubicon assurances with regard to questions about health and safety, social, environmental and wildlife concerns, as well as agreeing to provide Lubicon people with economic opportunities resulting from pipeline activities. As a starting point, they asked TransCanada "to recognize that the Lubicon Nation has unceded title to the lands through which the pipeline is expected to run" and that TransCanada should acknowledge and recognize "that the Lubicon Lake Indian Nation has never signed a treaty with the Government of Canada ceding Lubicon rights". The letter further asked TransCanada to acknowledge and recognize that the Lubicon had "unsettled aboriginal land rights over part of the route for the proposed North Central Corridor Pipeline Project" and that the company should acknowledge and recognize "that there is an on-going dispute between the Lubicon Lake Indian Nation and the

Government of Alberta over who exercises rightful regulatory authority" over the territory in which the pipeline project is being constructed.

In a response dated 15 November, and addressed to Chief Bernard Ominayak, not to the councillors who had written and signed the letter, TransCanada affirmed that it was open to hearing and addressing project-specific matters related to the pipeline, as well as supporting the resolution of Lubicon issues with the provincial and federal governments. However, TransCanada pointed out that as a public utility the company had "an obligation to build this facility in a timely manner and is therefore unable to further suspend its activities....We therefore notify you that we plan to recommence suspended project activities on Monday, November 17, 2008."[54]

The argument that projects should proceed in the public interest is being used increasingly to approve development projects in Alberta without public consultation. Northern Alberta forms part of what Coates and Morrison (1992) have called "the forgotten North," the vast area of sub-Arctic boreal Canada that falls within the provinces south of latitude 60° and is sometimes known as Canada's Middle North. Here, the process of industrial development, and its accompanying transformation of the environment, has occurred during a series of booms that have taken place more or less since the 1940s, a period characterized by Coates and Morrison as one of "turmoil and upheaval" in which treaties acted "as a kind of buffer for the Native people, giving them at least a measure of protection against exploitation" (ibid.: 84). In the case of the Lubicon Lake Indian Nation, no such buffer is in place and, in areas where treaty rights are evoked, such as the Woodland Cree community of Chipewyan Lake a little south of Lubicon territory, where recent seismic activity has been extensive, oil exploration and development and the absence of consultation may result in the de facto extinguishment of treaty rights if people "are denied meaningful participation in and control over the area's exploitation" (Timoney 2008: 2).

The community of Chipewyan Lake is faced with the development of deep-lying bitumen resources, yet oil companies operating in the area have provided local people with relatively little information and they are seeking answers to questions of concern with regard to disturbance to wildlife and habitat, air pollution and water quality. A collaborative project currently underway between researchers at the University of Alberta and the Woodland Cree community of Chipewyan Lake is currently exploring the tensions and contradictions that shape the decision-making abilities and processes of the community and other neighbouring Aboriginal communities confronted with energy exploration and development on their lands and is investigating the implications of oil development for community health, education and cultural sustainability. The community of Chipewyan Lake has already experienced significant energy development activity on its traditional lands, primarily from natural gas, along with the construction of pipelines and roads that cut through the boreal forest. Now, faced with the possibility of *in situ* bitumen development as close as 5 km from the village, residents of Chipewyan Lake are worried that this will forever transform their lives and lands. Their concerns are deepened because they argue that the project proponents have failed to provide the community with sufficient information that will allow them a basis for informed decision-making.[55]

The Northern Gateway Pipeline Project

The growth of Alberta's oilsands is not only linked to the development of new pipeline infrastructure to bring Arctic gas to northern Alberta, it depends on energy companies being able to send the results of oilsands production to existing and new markets. In 1998 Enbridge, one of North America's largest energy suppliers and gas distribution companies, began a feasibility study for a new pipeline to transport as much as 700,000 barrels of oilsands crude each day to the British Columbia port of Kitimat on Canada's west coast. The Gateway Pipeline Project would see the construction of a pipeline originating in Alberta, in Bruderheim just north of Edmonton, crossing north-west Alberta and northern British Columbia, and terminating some 1,170 km later at the new marine terminal, where oil would be shipped to markets in the U.S. and Asian Pacific Rim countries. The project, however, would see a second condensate gas pipeline transporting hydrocarbons back to Alberta from various sources around the world (and brought in by tankers to Kitimat) to contribute to bitumen production. Enbridge sees the route as passing mainly across Crown lands in both provinces.

The Canadian government has appointed a three-member Joint Review Panel to conduct the environmental assessment for the pipelines. The Canadian Environmental Assessment Agency (CEAA) and the National Energy Board announced the mandate and terms of reference for the panel in early December 2009. The panel will consider whether the project is likely to cause significant adverse environmental effects and if it is in the public interest. After conclusion of the review process, the panel will prepare a report setting out conclusions and recommendations and, following some government response to the report, the panel will issue a decision. While the panel will look at environmental effects of the Gateway project, as well as safety, technical, engineering and economic aspects, the impacts of increased oilsands production (including pollution, CO_2 emissions and the contribution the oilsands industry makes to global climate change) will not be within its mandate. At the time of writing this book, and prior to initiating the public hearings process, the Joint Review Panel was seeking comments from the public on the draft list of issues for the review, any additional information which the project proponents should be required to file, and the locations of the oral hearings.

Subsection 6.5 of the "Agreement between the National Energy Board and the Minister of the Environment concerning the Joint Review of the Northern Gateway Pipeline Project" states that, "In order that the Panel may be fully informed about the potential impacts of the project on Aboriginal rights and interests, the Panel will require the proponent to provide evidence regarding the concerns of Aboriginal groups, and will also carefully consider all evidence provided in this regard by Aboriginal peoples, other participants, federal authorities and provincial departments." Further, Subsection 8.1 discusses the nature of Aboriginal consultation, stating that, "In addition to Subsection 6.5, the Panel will receive information from Aboriginal peoples related to the nature and scope of potential or established Aboriginal and treaty rights that may be affected by the project and the impacts or infringements that the project may have on potential or established Aboriginal and treaty rights. The Panel may include in its report recommendations for appropriate measures to avoid or mitigate potential adverse impacts or infringements on Aboriginal and treaty rights and interests."[56] The Joint Review Panel's terms of reference state that the panel will take note of information provided by Aboriginal peoples regarding any ways the project may affect existing treaty rights. However, it has been claimed that the Joint Review Panel process and the framework for Aboriginal consultation do not accommodate the governance, management and decision-making rights of First Nations that are inherent in their Aboriginal Title.[57]

In a recent report for the Pembina Institute, Levy (2009) has argued that the health and abundance of salmon is crucial to the health of the environment in northern British Columbia as well as being of continued importance for the economies of many First Nations. The greatest concerns are the risks to salmon and freshwater habitat that would come from leaks and ruptures. Communities and First Nations along the proposed route have expressed concern at the risks posed to northern watersheds and other ecosystems and over the lack of consultation about the environmental review process. The Gitga'at First Nation, for example, has argued that Canada is in ongoing breach of its legal obligations to them in a number of ways, including the failure to recognize and affirm the Gitga'at First Nation's Aboriginal governance rights in its decision-making process in regard to the proposed Gateway project generally and in the choice and design of an environmental assessment process specifically; failing to seek to advance the reconciliation of the Gitga'at First Nation's Aboriginal governance rights with its own governance interests in its decision-making process in regard to the proposed project generally and in the choice and design of an environmental assessment process specifically; failing to acknowledge, consider or discuss with the Gitga'at First Nation the scope of its consultation obligation and how the scope of its obligation would be best reflected in the overall consultation process.[58]

In March 2010, Coastal First Nations, an alliance of nine First Nations on British Columbia's north and central coast, and on Haida Gwaii (the Queen Charlotte Islands), used the occasion of the 21st anniversary of the *Exxon Valdez* oil spill to announce the opposition of First Nations to the Northern Gateway Project. Gerard Amos, the alliance's director, said, "We will protect ourselves and the interests of future generations with everything we have because one major oil spill on the coast of British Columbia will wipe us out. This bountiful and globally significant coastline cannot bear an oil spill. This is where Enbridge hits a wall." A further declaration from the governments of the Coastal First Nations makes their position clear: "…in upholding our ancestral laws, rights and responsibilities, we declare that oil tankers carrying crude oil from the Alberta Tar Sands will not be allowed to transit our lands and waters." [59]

Essentially, the Coastal First Nations have asserted their right of ownership and control over their lands and waters as recognized in international law. In a legal comment on the declaration, Vancouver-based West Coast Environmental Law argued that the Coastal First Nations had the right to issue a ban on crude oil tankers in their coastal waters based on their own ancestral laws, and that a federal government decision to allow the Northern Gateway Pipeline Project and associated tanker traffic would infringe on the constitutionally-protected rights and titles of First Nations and breach Canada's obligations in international law.[60] Michael Ignatieff, leader of the federal Liberal Party, also announced that a Liberal government would formalize a moratorium on tanker traffic in northern British Columbia waters, prompting environmentalists to comment that this would be enough to kill the project.[61] Many of the First Nations of the Coastal First Nations alliance are in the midst of treaty negotiations or have filed writs claiming Aboriginal title. Together with opposition from political leaders, environmentalists and other concerned citizens, the Northern Gateway Pipeline Project faces the possibility of litigation from one or several First Nations. It could be a long drawn-out process, one that could be more contentious and controversial than the Mackenzie Gas Project.

CHAPTER VII

CONCLUSIONS:
COMMUNICATION, LEGITIMACY AND DIALOGUE

As the regulatory and public hearings process for the Northern Gateway Pipeline Project enters its initial phase, as First Nations and other residents of Yukon engage in discussion about how they need to prepare for what some see as the inevitable construction of the Alaska Highway Gas Pipeline, and as decisions are awaited on final regulatory approvals for the Mackenzie Gas Project, what lessons can proponents of large-scale projects in northern Canada and elsewhere in the circumpolar North, as well as communities faced with the development of those projects, learn from the diversity of approaches to proposing, evaluating, approving and regulating oil and gas development? The Mackenzie Gas Project, for instance, has for many (from industry, business and government) come to represent the inefficiency of federal environmental assessment rather than perhaps a model for future evaluation of megaprojects[62] but, for others, mainly those Aboriginal communities and other northern residents in the NWT (as well as Canadians living in other territories and provinces), environmental groups and citizen action groups opposed to the pipeline, or at least still uncertain about its benefits and impacts, the delays in the process have been welcomed.

In 1981, the Council for Yukon Indians sent a team on a study tour to Alaska to learn about the social and economic results of the Alaska Native Claims Settlement Act ten years after its implementation. The tour took place just as Yukon First Nations were preparing for negotiations in advance of land claims. The Alaska Study Tour Team wrote a report in which they were critical of the corporate nature of ANCSA and the alienation of people from the land, which they saw as a direct consequence of the negotiation and settlement process. Although the authors expressed solidarity with Alaska Native people, their criticism centred on the organizational structures and legal regimes within which people now have to move around and negotiate. "One strong impression we were left with," they wrote, "was that many Alaskan Natives, particularly the city-based managers and professionals, have forgotten who they are. They adopt organizational goals which are not different from non-native ones. They are distant from their origins." (Council for Yukon Indians 1981: 23). The tour and report influenced the way Yukon First Nations pondered the nature of what kind of land claim they would press for. For many, Alaska certainly did not provide a model and today, in stark contrast to the situation across the international border to the west, they have a right to participate in regulatory decision-making processes in ways that Alaska Natives cannot (Roddick 2006). This points to the importance of knowing and learning about precedent, of understanding the situations and experiences of indigenous peoples elsewhere in the circumpolar North, but also elsewhere in the world, of reflecting on them and using those situations and experiences to inform the choices and decisions people faced with development have to make.

Just as the Council for Yukon Indians looked to understand the situation in Alaska prior to positioning itself in relation to the federal government, so the Alaska Highway Aboriginal Pipeline Coalition has pointed to the importance of understanding how the Aboriginal Pipeline Group works for Aboriginal people in the Northwest Territories, as well as the significance of the Dene Tha' case in northern Alberta. Fogarassy (2007) shows that, as a result of the Dene Tha' case, the law of Canada is that Aboriginal people who assert Aboriginal (or treaty) rights that may potentially be affected by a project which is dependent on Crown decisions and authorization must be consulted very early in the project planning process. The judgment, he argues, means that consultation must be meaningful. Not to engage in consultation, not to hear the voices and concerns of Aboriginal people endangers the timelines, viability, validity and legitimacy of the project. It is the Dene Tha' case that has set a precedent for how things should and must be done—it has increased the bargaining power of Aboriginal people faced with energy development (Fogarassy ibid.). The duty to consult obliges the Crown to avoid decisions which may result in irreparable harm to the lands and waters of Aboriginal peoples—such decisions may be challenged in court.[63]

In hearings for the Mackenzie Gas Project, as well as in discussions over pipeline development in Yukon or British Columbia, or in the expression of local anxieties over bitumen extraction in northern Alberta, Aboriginal people stress the importance of discussing the sustainability of resource development in relation to the sustainability of indigenous and local livelihoods, societies and cultures (at all levels, from that of the household, through to community, region and society), and the health of the environment. While oil and gas activities impact on traditional resource use practices and local livelihoods, it is important not to forget that these social and economic impacts cannot and must not be considered and analyzed in isolation from the impacts of other contemporary changes, such as rapid social and cultural change and climate change. The Arctic Climate Impact Assessment, released by the Arctic Council in 2005, shows that, in the face of current and projected climate change, the Arctic's indigenous peoples are facing special and unique challenges and their abilities to continue to harvest wildlife and access food resources are already being put to the test. Being able to access traditional food resources and ensuring food security will be a major challenge in an Arctic affected increasingly by climate change and global processes. For many, the prospect of environmental damage resulting from energy development only exacerbates this challenge.

The low diversity of economic options in much of the Arctic already renders the region vulnerable to changes in locally available resource bases and in global economic trends and markets. This goes some way to explaining the northern phenomenon of cycles of boom and bust which characterize so many parts of the Arctic and sub-Arctic (Chapin et al. 2004). The enhancement of local and regional economic diversification is a critically-important step towards increasing the resilience of Arctic communities and economies. Yet some of the main concerns associated with both climate change and major new oil and gas development activities in the circumpolar North revolve around safety issues, the increased risk of accidents, the increasing difficulty in gaining access to food resources, and the contamination of ecosystems and wildlife with pollutants. Increased ground temperatures resulting from rising air temperatures and snow depth, for instance, will increase the likelihood of the transport of contaminants and, when it comes to pipeline projects a number of geotechnical climate change issues need to be addressed when constructing pipelines in permafrost zones (Furgal and Prowse et al. 2007: 79-80). Key questions for informing further research and policy discussion include:

- How will the ability to be resilient to climate change be challenged or compromised by oil and gas activities?
- How can traditional and local knowledge be used most effectively in monitoring biodiversity at the circumpolar level to detect the impacts of oil and gas activities, and to understand these changes within the context of broader global changes?
- How can local communities respond effectively to such impacts and changes?
- What kinds of threats and problems will occur as a result of environmental disturbance in the consumption of traditional and local foods?
- Furthermore, what kind of communication is already taking place between communities, developers and policy makers with respect to the impacts of oil and gas activities on traditional resource use activities? This is especially important to consider with respect to dietary advice to Arctic peoples so that they can make informed choices about the foods they eat in situations of environmental disturbance.

Migrant labour is one of the most distinctive features of oil and gas development and its impacts are felt during both construction and operational phases of development. This has attracted considerable attention from social scientists researching the impacts of migration and demographic change on oil-dependent parts of Scotland, Norway and Newfoundland (e.g. Moore 1981), but is less-well chronicled, if at all, in the Arctic. A major criticism of the level and extent of migrant labour in the north of Scotland during the oil boom of the late 1970s and early 1980s, for instance, was that it reduced job opportunities for the local unemployed, and had disruptive effects on local communities, for example in the availability of housing, rents, house prices and related social problems (Mackay 1981, Jedrej and Nuttall 1996). Skill requirements are an obvious and major factor in deciding the balance of local and migrant labour (Mackay ibid.). Commonly, areas affected by migrant labour have not had an indigenous labour pool, so temporary and permanent immigration has been necessary.

Unlike mining, oil and gas discoveries do not usually lead to the establishment and construction of towns. However, nearby cities and towns tend to develop into bases for development and centres within which the oil and gas industry clusters. In some cases, communities grow to three times their size due to, mainly male, immigrants. The heavy influx of workers is apparent for many years after exploration or construction has ended and can create long-term social, political and economic changes in local communities. There are situations where this may seem beneficial to the fortunes of communities, as is being reported in Hammerfest in northern Norway since the construction of StatoilHydro's LNG plant on the nearby island of Melkøya, to which gas from the Snøvhit field in the Barents Sea is sent through an underwater pipeline, yet concerns remain about both short-term and long-term impacts of in-migration. A number of key questions need to drive future research, stakeholder dialogue and policy discussion, including:

- What implications follow from migrant labour patterns in terms if reducing job opportunities for local people? Indeed, what is the local labour availability in areas affected by hydrocarbon development?
- What kinds of disruptive effects does migrant labour have on local housing markets?
- What kinds of social problems follow as a result of an influx of migrant workers into small communities?

- What kinds of impacts and consequences are there as a result of different types of development (e.g. temporary construction projects and the permanent labour force)?
- Furthermore, what kinds of issues and concerns arise for members of local communities who themselves become migrant workers and travel away from home regularly to work in the oil and gas industries? What are the consequences for local social and economic dynamics?

Related to this are issues of family, household, community and the nature of work. Little research has been carried out on the complex interrelationships between employment, workplace policies and family and household life that have already occurred, and are likely to occur, within the oil and gas industries. There is urgent need to understand the kinds of dramatic changes which occur in local and regional labour market structures, local culture and community life as a consequence of oil and gas development. For example,

- How does/will the changing nature of work as a result of oil and gas development affect households, families and children, and what opportunities and constraints does/will it present for them?
- How do children and young people experience and understand the working patterns of their parents and adult family members?
- What kinds of impacts do/will wider labour market processes have on families, households and communities?
- How do children perceive oil and gas work and the oil and gas industry?
- To what extent do children's views influence, directly or indirectly, parental work decisions?
- What occupational health and employment policies related to work-family issues are in place across the oil and gas industry?
- What are/will be the policy implications of taking the expressed views, needs and interests of children and parents into account with regard to work-family issues?

As this book has discussed, oil and gas development activities in the Arctic are critically important to indigenous peoples, who are increasingly concerned about the interest of industry and national governments, and the far-reaching impact of the world market on their northern homelands. As Nellie Cournoyea put it at one of the Mackenzie Gas Project hearings,

Our caution is fuelled by the understanding that there will be unavoidable social impacts from this and other hydrocarbon projects in the years ahead, and also by the recognition that we must be eternally vigilant in ensuring our natural environment is not diminished by the very forces that feed our economic well-being.[64]

Fondahl and Sirina (2006a, 2006b) discuss how, for much of the late 20th century, hydrocarbon development in Russia took place on native lands, largely in the western Siberian oil fields on Khanty and Mansi traditional territory and the north-western gas fields on Nenets homelands. More recently, controversy over the exploitation of Sakhalin Island's oil reserves has challenged the territorial rights of the Nivkhi, Evenki and Uilta. While Fondahl and Sirina discuss whether Russia's indigenous peoples have received some benefits from oil and gas development, they argue that generally the costs outweigh whatever positive experiences can be chronicled. Oil and gas extraction has removed significant lands and territories from indigenous use, and the transport of hydrocarbons across indigenous lands also impacts on indigenous activities. Pipelines dissect reindeer pastures, affect the movement of animals hunted for food, and hinder traditional access enjoyed by indigenous peoples to hunting and herding grounds. "Pipeline failures and leaks," they say, are

an all too common occurrence in Russia, cause pollution to land and water sources, and thus affect traditional indigenous land- (and water-) based activities such as hunting, gathering and reindeer herding. Pipeline construction brings influxes of outsiders into the area, which has both benefits and disadvantages. Apart from the environmental degradation problems posed by leaks, the

construction phase may affect animal habitat and culturally im-
portant sites. Once in place the pipeline may disrupt migration
paths of terrestrial animals, including domesticated reindeer and
wild game.

(Fondahl and Sirina 2006a: 58)

Murashko (2008) points to the lack of public participation in regulatory processes concerned with large-scale development in Russia,
arguing that procedures have not been established to ensure the
participation in referendums and public hearings of those peoples
whose traditional lands will be impacted by resource development
projects. Public monitoring by indigenous peoples, she argues, is
replaced with superficial project presentations for citizens at meetings in cities and towns far from the actual project site. Povoroznyuk (2006) argues that the severe social and economic problems
resulting from industrialization—specifically oil and gas activities—suggest that there is a need for stronger forms of government
regulation, national support for indigenous peoples, and greater
involvement of indigenous peoples in decision-making processes.

In 2005, public hearings were organized for the Eastern Siberia-Pacific Ocean oil pipeline, which was originally proposed by
Yukos and is currently being constructed by Transneft and, when
finished, will stretch 5,000 km across eastern Siberia, transect indigenous lands and pump 1.6 million barrels of oil a day to Kozmino
Bay near the port of Nakhodka on the coast of the Sea of Japan (see
also, Fondahl and Sirina 2006a, 2006b, Sirina 2009). China was the
original main market, but interest in Siberian oil has also been expressed by Japan, India and the United States. Povoroznyuk (ibid.)
describes how the structured formality of the public hearings as
well as the unhampered positive coverage of the pipeline project
by the majority of the regional and national media sources, did
not leave much choice or possibility for an adequate evaluation of
the social and economic impacts that this development will have
on indigenous communities. Fondahl and Sirina (2006a: 65) have
also described the frustration expressed by indigenous people because they were unable to express their views and have a voice in
the project: "Many are not categorically opposed to the project but

rather want to ensure that ecological safeguards are in place, and that they benefit from the construction of such a project through their homelands, whether through compensation payments or through employment opportunities."

The project courted controversy and was opposed by environmentalists as it was originally planned that the pipeline would pass close to the southern shore of Lake Baikal (which is the world's oldest lake and was declared a UNESCO World Heritage Site in 1996). Transneft was also accused of illegal tree felling in the area prior to any approval for the route. The Lake Baikal route was subsequently abandoned in favour of one to the north of the lake. Concern was also expressed over the original site for the terminal at Perevoznaya on Amur Bay. Environmentalists worried about the impact on terrestrial and marine wildlife—for example, the coastline of Amur Bay is the habitat of the rare and endangered Amur leopard, of which only 30 or 40 are left in the wild (including ten of that figure in China), as well as being an area of a wealth of biodiversity as a result of the meeting of boreal and subtropical waters. Local people also depend on the coastal waters for fishing. Environmentalists began calling Perevoznaya Bay "Siberia's Prince William Sound" evoking memories of the sinking of the *Exxon Valdez* in 1989 and warning of possible future oilspill catastrophes (Brooke 2005). The eventual decision to route the pipeline to Nakhodka was probably determined by the practicalities of economics rather than by concern for the environment, as it provides a strategic location for a distribution centre to markets other than China. As Roon (2006) and Stammler and Wilson (2006) also show for Sakhalin Island, the Sakhalin-1 and Sakhalin-2 projects have drawn attention to the debate around the assessment of industrial impacts on local (indigenous and non-indigenous) communities in the Russian North.

Despite such situations, it is not a simple issue, however, of traditional cultures facing the onslaught of change and disruption brought about by industry. As this book has discussed, indigenous business and community leaders see and reap benefits from such development and, indeed, companies owned by indigenous people are involved in the energy sector. This is particularly the case in Alaska and northern Canada. In western Siberia, indigenous groups have a history of close relations with the government and the largely state-run oil companies which has resulted in a situation of coexistence (Stammler and Forbes 2006). In the Russian Far East, concern over oil and gas activity has led to Sakhalin's indigenous groups forging alliances with national and international NGOs and Russian political parties (Roon ibid., Stammler and Wilson ibid.) and Sakhalin's indigenous leadership has become increasingly politicized and influential in the ways it deals with and negotiates with the oil companies operating on the island. In Greenland, as the Danish North Atlantic territory acquires a greater degree of political autonomy, the Inuit-led government is hoping that oil, gas and mining will become major drivers of the economy, with some political leaders expressing ambitions that resource extraction will generate revenues that will help Greenland achieve full independence from Denmark (Nuttall 2008b). Greenland's Premier Kuupik Kleist recently remarked that

The living resources such as fish and marine mammals are important sources of income, which will and should remain so for generations to come. However, despite innovative product development, climate change may affect the reliability of the catch. Income from the fisheries and marine mammals are not large enough to sustain the welfare levels that the people of Kalaallit Nunaat needs, deserves and asks for.

We are therefore very keen to develop and diversify our sources of income. In an age of rapid global warming, we also hope that the sheep farmers of the southernmost municipality of Greenland will become important contributors to our domestic consumption of meat and vegetables. Tourism is a sector from which we have great expectations and of which we expect increased earnings. However, the exploitation of our enormous riches in oil and mineral resources is indisputably the most promising and real potential for a greater degree of economic self sufficiency – at a scale that will secure Greenland's economy base and our future livelihoods.[65]

Nevertheless, despite the promise of lucrative income and the sustainable development of communities, anxiety persists among Arctic residents about the cumulative effects of historical and proposed activities on resources, lands, waters and cultures (NRC 2003). For example, in Yamal where, despite a situation of coexistence, the fact remains that local residents still have little influence over resource extraction and change has hit a crisis point, challenging the ability of local residents to adapt to the pace of oil and gas development (Krupnik 2000). Who has rights to land—indeed, whose land it is—is at the heart of much of the conflict, misunderstanding and concern associated with energy development. At the Joint Review Panel hearing for the Mackenzie Gas Project in Deline in Canada's Northwest Territories, Caroline Yukon was just one of many who expressed her concern about the future of the land:

We love our land so much. We don't want anybody to take it away from us. It's ours, and we have to protect our land. No matter what anybody says or what anybody is going to say, they're going to do something about it. We have to protect our land no matter what... Once we lose our land, we're never, ever going to replace it again. You have to think about that.[66]

Knowing how to engage in constructive and effective dialogue with industry—or rather, ensuring the recognition of the right to engage in constructive and effective dialogue—remains a challenge for indigenous peoples and this challenge increasingly influences the relations between oil and gas companies and local residents throughout the North. Although they face and experience similar kinds of impacts, consequences and opportunities, communities across the Arctic are and will be affected by oil and gas development in markedly and profoundly different ways. Local experiences and responses to this development may not be universal, but the nature of the experience of energy development becomes of significant interest to indigenous peoples. It becomes pertinent to look to the past and to explore the experiences people have had with oil and gas development. How did, for example, local and regional authorities and communities anticipate, prepare for, deal with and benefit (or not) from the arrival and presence of industry? How were informed decisions made, and who were the decision-makers? How did processes of engagement between local and regional authorities and communities and industry proceed? What was the nature of discussions on the regulatory framework, environmental and social effects, the public review process, training and employment opportunities? And, as people pondered the changes about to occur and imagined themselves into a future dependent on the oil and gas industry, what does the present look like now that the future has been reached?

The rights of indigenous peoples to cultural sustainability, to the protection, conservation and management of their natural resources and to determine their own development are guaranteed within international human rights law, even if these rights are not necessarily always recognized by nation-states with indigenous peoples.[67] In Canada, Aboriginal and treaty rights are recognized and affirmed in Canada's *Constitution Act*, 1982, but how indigenous people in northern Canada are now positioned in relation to energy development in terms of consultation, active participation and economic benefits is the result of a process of historic treaty-making and more recent negotiations over comprehensive land claims. This places an obligation on government and public institutions to acknowledge and respect these rights, including the right to maintain identity, the right to lands and resources, and the right to continue cultural practices.

Fjellheim and Henriksen (2006) have put forward an indigenous perspective on the international human rights protection accorded to indigenous lands and resource rights, with particular reference to oil and gas exploration. They argue that indigenous peoples have been, and in many cases still are, deprived of their human rights and that this has resulted in a situation whereby they experience the dispossession of their lands, territories and resources. This denial of rights and a fundamental recognition of indigenous peoples as distinct peoples has, in many cases, prevented them from exercising their right to development in accordance with their needs and interests. In the case of the oil and gas industry, there has been a history of widespread failure around the world to recognize and respect indigenous peoples' use, occupancy and ownership of their traditional lands, territories and resources. Saami in northern Norway, for example, have expressed their rights to receive part of the revenues from oil and gas production in the Barents Sea. In arguing their case, they have looked to the Canadian North and to how the Inuvialuit of the western Arctic have managed their participation in the Mackenzie Gas Project.

What kinds of consequences do/will oil and gas development activities have for self-government and autonomy in the Arctic? Research in areas that have experienced extensive oil and gas development has shown that the large-unit size and sheer scale of most oil and gas-related development can actually increase the *dependence* of local communities and regions upon national governments and transnational corporations. The financial and employment benefits that may flow to local communities as a result of oil and gas development may be countered by increasing dependence on national government for the provision of infrastructure, environmental assessments, anti-pollution measures, occupational health and safety policy, and for policy responses to the uncertainties and fluctuations inherent in the global energy economy. As a result of oil and gas development in the Arctic, will local communities, regions and local authorities be constrained in their abilities to regulate local economic and political life?

Non-renewable resource extraction alone cannot support a strong foundation for social and economic sustainability. As a concept, effective sustainable development must encompass more than just economic growth. It must be concerned with broader human dimensions, with well-being at the levels of individual, family, household and community, as well as with social equity and human responsibility towards present and future generations. An approach focused on sustainable livelihoods emphasizes concern for people, ways of life and local forms of development which are appropriate to local social, cultural and economic conditions, needs and circumstances, and which must be worked out, defined and implemented within the context of people's relationships to one another and to the environment.

At the same time, as is apparent when one looks at the multifaceted nature of development projects and the social, cultural, economic, political and institutional contexts in which they occur, there are a diversity of understandings and numerous definitions of sustainable development throughout the Arctic and sub-Arctic, as well as multidimensional forms of interaction and dialogue between governments, industry, local communities and other stakeholders. Often, these are expressed in institutional and public forms of discourse that are at odds with one another. Furthermore, the very legitimacy of claims, positions and knowledge comes under scrutiny. In developing his theory of communicative action, Habermas (1984, 1990) argued for the importance of discursive democratic procedures in public spheres—ways of allowing effective participation and working towards consensual understanding rather than compromise and coercion. A fundamental starting point for processes of environmental impact assessment and for strategies of corporate social responsibility—and for their eventual social legitimation— is for industry and government to understand and recognize this diversity and to be ready to hear a plurality of voices, to understand local contexts, to have an awareness of the histories of societies and local communities, to understand the importance and value of lands and waters for local people, and be open to understanding cultural practices and local expressions of human-environment relations. As Thomas Berger discovered in the 1970s, sitting and listening to people articulate their concerns, fears, hopes, ambitions and dreams, and listening to them tell stories, is a crucial part of environmental impact assessment.

Notes

1   From 23-25 March 2009, representatives of indigenous peoples from around the world met in Manila in the Philippines to participate in the International Conference on Extractive Industries and Indigenous Peoples. From the Arctic to the tropics, from the rainforests of South America and South East Asia, and from the deserts and arid lands of Australia and Africa, the event highlighted indigenous concerns over oil, gas, and mining development activities. The resulting declaration from the conference—the Manila Declaration—emphasized that "we have suffered disproportionately from the impact of extractive industries as our territories are home to over sixty percent of the world's most coveted mineral resources. This has resulted in many problems to our peoples, as it has attracted extractive industry corporations to unsustainably exploit our lands, territories and recourses without our consent. This exploitation has led to the worst forms of, environmental degradation, human rights violations and land dispossession and is contributing to climate change." The full text of the declaration can be accessed at the Tebtebba website: http://www.tebtebba.org/index.php?option=com_docman&task=doc_download&gid=358&Itemid=27

2   **Bird et al., 2008:** *Circum-Arctic Resource Appraisal; estimates of undiscovered oil and gas north of the Arctic Circle*, U.S. Geological Survey Fact Sheet 2008-3049, 4 p. [http://pubs.usgs.gov/fs/2008/3049/].

3   Pratt (2001: 21) argues that the United States and Canada are already experiencing such problems: "Their best conventional oil supplies are depleting rapidly, gas is abundant but no longer cheap, while most of the remaining undiscovered reserves of oil and gas on the continent are located in difficult, hotly-disputed wilderness areas: the eastern Gulf Coast, the Rocky Mountains, the Alaskan wildlife refuge, the Alberta and B.C. Foothills, and the West coast onshore and the Beaufort Sea."

4   A number of recent books deal with Arctic sovereignty in the context of ownership of territory and resources, such as work by Byers (2009), Howard (2009) and Sale and Potapov (2010).

5   See Byers (2009) for a fuller account.

6   A number of recent reports provide accounts of the social, economic and political situations of the indigenous peoples of the circumpolar North. See, for example, ACIA (2005), and AHDR (2004).

7   The full text of the "Circumpolar Inuit Declaration on Arctic Sovereignty" can be accessed from the website of Inuit Tapiriit Kanatami: http://www.itk.ca/circumpolar-inuit-declaration-arctic-sovereignty. Article 3.6 of the Declaration asserts: "As states increasingly focus on the Arctic and its resources, and as climate change continues to create easier access to the Arctic, Inuit inclusion as active partners is central to all national and international deliberations on Arctic sovereignty and related questions, such as who owns the Arctic, who has the right to traverse the Arctic, who has the right to develop the Arctic, and who will be responsible for the social and environmental impacts increasingly facing the Arctic. We have unique knowledge and experience to bring to these deliberations. The inclusion of Inuit as active partners in all future deliberations on

Arctic sovereignty will benefit both the Inuit community and the international community."

8　In a response to the National Energy Board announcement, the Canadian Association of Petroleum Producers said that: "We believe this review will provide a forum for a thoughtful dialogue on these issues and for consideration of lessons learned from the Deepwater Horizon incident in the Gulf of Mexico.Our industry recognizes that we must take the time to assess the lessons learned and theimplications for Arctic drilling." (Available at the NEB public registry, https://www.neb-one.gc.ca/ll-eng/livelink.exe/fetch/2000/90463/621169/622976/A1T3H3_-_Letter_of_Comment.pdf?nodeid=623072&vernum=0). The Inuvialuit Game Council also responded and made the interesting observation that "as a result of the tragedy in the Gulf of Mexico we have been examining drilling statistics worldwide, and were surprised to discover that many — if not most — blowouts occur on land. Therefore, it is perhaps prudent to reconsider the risks of all exploratory and development drilling, both offshore and onshore. Further to this, with a rapidly changing climate it is becoming less clear what threats are likely to compound the risks already involved in exploratory and development drilling." (Available at the NEB public registry, https://www.neb-one.gc.ca/ll-eng/livelink.exe/fetch/2000/90463/621169/624556/A1T5A8_-_Letter.pdf?nodeid=624477&vernum=0.

9　Kleveman (2003: 7-8).

10　In Canada, Aboriginal people is the official term used in the *Constitution Act, 1982* for the country's indigenous Inuit, Indian (First Nations) and Métis peoples. I use Aboriginal in this book according to conventional and accepted usage, but I also use the term indigenous to refer to Aboriginal people both in Canada and globally.

11　For some specific case studies, see Nuttall and Wessendorf (2006), Mikkelsen and Langehelle (2008), and the Arctic Council's Oil and Gas Assessment at http://www.amap.no/oga

12　Inuit Circumpolar Council 2010 "Nuuk Declaration" (http://inuitcircumpolar.com/index.php?ID=435&Lang=En).

13　The Arctic Leaders' Summit, held in Hay River on 11-12 December 2005, provided an opportunity for many different voices to come together to look at the issue of climate change and oil and gas development in the North. The discussions focused on community participation, health, youth and the International Polar Year (which later ran from 2007-2009). At the close of the summit, leaders and participants developed the Katlodeeche Plan of Action which, amongst other things, called for inclusion within current, national and bilateral and circumpolar cooperative activities, including the Canada-Russia Intergovernmental Economic Commission, provision for the sharing of information between countries, their agents and regulatory authorities, on best practices in the oil and gas development and exploration industry, as well as key principles for their implementation, with specific reference to the access and benefit-sharing arrangements for indigenous and local communities. See also, Arctic Athabaskan Council (2006), Tesar (2006).

14　**Mackenzie, Alexander, 1911:** *Voyages from Montreal through the Continent of North America to the Frozen and Pacific Oceans* 2 vols. Courier Press: Toronto. Mackenzie's journals were originally published in 1801.

15 Rohmer's book was a call for Canada to formulate national priorities and policies on energy research and development in the North. Worried by increasing interest in Canada's natural resources from the United States, he argued that Canadian ownership and control of oil and gas would be threatened by American corporations: "Why are the drilling rigs working so feverishly in the remote, hostile Canadian Arctic? They search because mankind's civilization is today totally dependent upon natural gas and oil, the elusive fossil fuels secreted millions of years ago under layer upon layer of rock. Their search is urgent because on the North American continent there is an increasing shortage of energy as the population demands outstrip the supplies" (Rohmer 1973: 8).

16 India has also expressed interest in making investments in Alberta's oilsands (Polczer 2010).

17 See Grace (2001) for a fascinating discussion of the ideas, representations and images of the North in the historical and cultural narratives of Canada.

18 For further information see:
http://www.grrb.nt.ca/traditionalknowledge.htm

19 National Energy Board. 2006a. *Mackenzie Gas Project* Volume 1, Hearing held at Inuvik (Northwest Territories), Wednesday 25 January 2006. Ottawa: International Reporting Inc.

20 Joint Review Panel for the Mackenzie Gas Project. 2006a. *Volume 2: Hearing held at Inuvik, NWT, 15th February 2006*. Ottawa: International Reporting Inc, p. 67

21 Kakfwi, S. 2001 "Maximizing Aboriginal benefits from resource development in the Northwest Territories", a presentation given at *First Nations Rights and Aboriginal Interests in Northern Pipeline Development Conference*, Calgary, Alberta, 18 June 2001.

22 Joint Review Panel for the Mackenzie Gas Project. 2006b. *Volume 2: Hearing held at Inuvik, NWT, 15th February 2006* Ottawa: International Reporting Inc, p. 21-22

23 Joint Review Panel for the Mackenzie Gas Project. 2006c. *Volume 4: Hearing Held at Fort McPherson, NWT, 17th February 2006*. Ottawa: International Reporting Inc., p.283.

24 Joint Review Panel for the Mackenzie Gas Project. 2006d. *Volume 27: Hearing Held at Pehdzeh Ki Dene Band Complex, Wrigley NWT, 11th May 2006*. Ottawa: International Reporting Inc., p. 2461.

25 Joint Review Panel for the Mackenzie Gas Project. 2006e. *Volume 4: Hearing Held at Fort McPherson, NWT, 17th February 2006*. Ottawa: International Reporting Inc., p.288.

26 Joint Review Panel for the Mackenzie Gas Project. 2006f. *Volume 4: Hearing Held at Fort McPherson, NWT, 17th February 2006*. Ottawa: International Reporting Inc., p.291.

27 Joint Review Panel for the Mackenzie Gas Project. 2006g. *Volume 4: Hearing Held at Fort McPherson, NWT, 17th February 2006*. Ottawa: International Reporting Inc., p.301.

28 "Arctic Indigenous Youth Alliance Submission" Public Registry for the Review of the Mackenzie Gas Project, http://www.ngps.nt.ca/registryDetail_e.asp?CategoryID=94.

29 Joint Review Panel for the Mackenzie Gas Project. 2006h. *Volume 3: Hearing held at Inuvik, NWT, 16th February 2006* Ottawa: International Reporting Inc, p. 113

30 http://www.dehchofirstnations.com/home.htm

31 Joint Review Panel for the Mackenzie Gas Project. 2006i. *Volume 26: Held at Fort Simpson, NWT, 10th May 2006*. Ottawa: International Reporting Inc. p.2368.

32 "Alberta band joins list of pipeline opponents". 18 May 2005 http://www.cbc.ca/canada/north/story/2005/05/18/dene-tha-pipeline-170512005.html.

33 "Alberta band loses bid for pipeline hearing delay". 9 January 2006 http://www.cbc.ca/canada/north/story/2006/01/09/dene-pipelin-060912006.html.

34 Jaremko, G. 2006. "Dene discuss pipeline under protest". *Edmonton Journal*, Friday 7 July, D1.

35 Joint Review Panel for the Mackenzie Gas Project. 2006j. *Volume 51: Held at Ulukhaktok, NWT, 8th September 2006*. Ottawa: International Reporting Inc., p.4975.

36 Joint Review Panel for the Mackenzie Gas Project. 2006k. *Volume 51: Held at Ulukhaktok, NWT, 8th September 2006*. Ottawa: International Reporting Inc., p.4972.

37 Joint Review Panel for the Mackenzie Gas Project. 2006l. *Volume 51: Held at Ulukhaktok, NWT, 8th September 2006*. Ottawa: International Reporting Inc., p.4973.

38 Stevenson, J. "For Devon, it's drill or forfeit Beaufort Sea offshore licence". *Edmonton Journal* 7 June, D7.

39 Dehcho First Nations, "Review of comments from Imperial Oil". Submission to the National Energy Board, 11 February 2010.

40 Dehcho First Nations Response to Joint Review Panel. Submission to the National Energy Board, 11 February 2010.

41 Sambaa K'e Dene Band Response to Joint Review Panel. Submission to the Government of Canada and the National Energy Board, 11 February 2010.

42 Carroll, J. 2007 "Costs may scuttle Mackenzie pipeline". *Edmonton Journal*, Thursday 31 May, F1.

43 The Foothills (TransCanada) natural gas transmission system carries natural gas for export from central Alberta to the U.S. border to serve markets in the U.S. Midwest, Pacific Northwest, California and Nevada. Foothills was built and expanded under the legislative framework of the Northern Pipeline Act (NPA). Any future expansions of the Foothills System would also be under the jurisdiction of the NPA, while ongoing operations are governed by the National Energy Board.

44 In learning how they can respond effectively to development, the sharing of experiences by indigenous peoples from different regions is critical for effective dialogue. During the Whitehorse regional leaders' workshop in June 2005, Chief Randy Mayo from Stevens Village, Alaska, and then AAC Alaska chairman, stated that it was important to learn about and share perspectives on cross-border issues since the Alaska-British Columbia-Yukon border was an imaginary political line that cut across hereditary cultural and tribal lines (Arctic Athabaskan Council 2005).

45 See the Alaska Highway Aboriginal Pipeline Coalition website, http://www.ahapc.ca

46 Energy Information Administration http://tonto.eia.doe.gov/state/state_energy_profiles.cfm?sid=AK

47 Presentation made to the Governor of Alaska's Alaska Highway Natural Gas Policy Council, 19 July 2001. Available at: http://www.arcticgaspipeline.com http://www.arcticgaspipeline.com/Reference/Documents&Presentations/A-Council/7-19-01BarrowCouncil/7-19-01ASRC-R_Glenn.doc

48 Gwich'in Steering Committee, www.gwichinsteeringcommittee.org

49 Gwich'in Niintsyaa (Resolution) http://gwichinsteeringcommittee.org/gwichinniintsyaa.html

50 Alaska Department of Environmental Conservation, North Slope Oil Spill Database, 2004.

51 Once completed, the Keystone pipeline would involve more steel pipe in the ground than the proposed Alaska Highway Gas Pipeline. By 2012, the pipeline is expected to move 910,000 barrels a day (O' Meara 2010).

52 Liquid toxic waste called tailings is a by-product of the energy extraction process. Tailings are stored in large ponds near oilsands mines. In 2010, Syncrude went on trial and faced both provincial and federal charges over the death of 1,600 ducks in its Aurora mine's tailing pond in 2008. The company was found guilty of failing to prevent their deaths and, while the federal prosecutor ruled out prison sentences for Syncrude executives, Syncrude was ordered to pay a penalty of Can$3 million.

53 "The Rights of the Lubicon Cree must be Protected". Joint Public Statement and Briefing by Alberta Federation of Labour, Amnesty International, Canadian Friends Service Committee (Quakers), Canadian Labour Congress, Cultural Survival, First Peoples Human Rights Coalition, Greenpeace Canada, Indigenous Environmental Network, IWGIA, KAIROS: Canadian Ecumenical Justice Initiatives, Public Service Alliance of Canada (PSAC), Tebtebba - Indigenous Peoples' International Centre for Policy Research and Education. The full statement can be accessed and read on the website of the International Work Group for Indigenous Affairs, at http://www.iwgia.org/sw40965.asp as well as on the website of Amnesty International at http://www.amnesty.ca/resource_centre/news/view.php?load=arcview&article=5239&c=Resource+Centre+News. The statement claimed that the situation of the Lubicon Cree "typifies challenges facing Indigenous Peoples in Canada" and also referred to the visit of the UN Special Rapporteur Miloon Kothari's visit to the Lubicon community in 2007 and his statement in Ottawa in which he described "appalling living conditions" and "the asphyxiation of livelihoods and traditional practices."

54 To view the correspondence see http://www.lubicon.org

55 Funded by the Social Sciences and Humanities Research Council of Canada (SSHRC) and running from 2009-2012, the project "Energy Development and Community Well-being in Northern Alberta" involves myself and colleagues Makere Harawira-Stewart, Clifford Cardinal and Kevin Timoney. The project aims are to a) document the environmental and ecological impacts of resource extraction and development within the traditional area of the northern Aboriginal community of Chipewyan Lake and b) monitor, record and document the impact on the community itself and its responses as it endeavors to maintain and protect its traditional subsistence lifestyle in the face of impending oil, gas and logging development. The project has been developed over a series of preliminary consultations held with the community association and explores avenues of concern highlighted by community members themselves. During community visits, people expressed concern that information about energy development near the community was not necessarily forthcoming from industry and government or that, at best, whatever information was being given to them was vague. The project is underpinned by community awareness of the impacts of resource extraction on the health and traditional lifestyles of Aboriginal communities in Northern Alberta, particularly those in Fort Chipewyan, Fort McMurray and Fort McKay.

56 *Joint Review Panel Agreement, 4 December 2009, available at the Canadian Environmental Assessment Agency website:* ttp://www.ceaa-acee.gc.ca/050/document-eng.cfm?document=39960

57 West Coast Environmental Law, http://wcel.org/resources/publication/legal-comment-coastal-first-nations-no-tankers-declaration, p.4.

58 *Gitga'at First Nation, 27 November 2009,* "Letter regarding the Draft Joint Review Panel Agreement and Terms of Reference", available at the Canadian Environmental Assessment Agency website: http://www.ceaa-acee.gc.ca/050/document-eng.cfm?document=39875

59 See the press release on the website of Coastal First Nations: http://coastalfirst-nations.ca/files/PDF/C35010032215240.pdf

60 West Coast Environmental Law, "Legal Comment on Coastal First Nations Declaration 'No Tar Sands tankers in our waters'", Vancouver, 24 March, 2010. The legal comment points out that: "Free, prior and informed consent is the international standard government consultation with First Nations on issues such as approval of the Enbridge Northern Gateway Pipeline project and related tanker traffic..... A decision by the federal government to approve the Enbridge Northern Gateway Pipeline project and related oil tanker traffic, in the absence of First Nations consent, would violate Canada's international legal obligations, and make Canada vulnerable to a human rights challenge in an international (e.g., UN Human Rights Committee) or regional (e.g., Inter-American Commission on Human Rights) forum. If a First Nation takes international legal action against Canada for such a decision, there is a significant risk that a finding would be made against Canada, attracting negative world attention and creating further uncertainty for the Enbridge project. In addition, international human rights bodies such as the Inter-American Commission can request Precautionary Measures, which may include a request that a project not proceed further until such time as the First Nation's petition can be heard and decided on its merits." http://wcel.org/resources/publication/legal-comment-coastal-first-nations-no-tankers-declaration, p.3)

61 "Liberals back B.C. tanker ban". *Edmonton Journal* 22 June 2010, p. E1.

62 At the Western Premiers' Conference in Vancouver in June 2010, the premiers of British Columbia, Alberta, Saskatchewan, Manitoba, Yukon, Northwest Territories and Nunavut called on the federal government in Ottawa to combine provincial, territorial and federal environmental assessments into one process. While the premiers expressed frustration at the length of time it can take for major development projects to get regulatory approval, environmentalists expressed concern that streamlining the process would mean less environmental rigour for projects (Crawford 2010).

63 West Coast Environmental Law, http://wcel.org/resources/publication/legal-comment-coastal-first-nations-no-tankers-declaration.

64 Joint Review Panel for the Mackenzie Gas Project. 2006m. *Volume 1: Hearing held at Inuvik, NWT, 14th February 2006.* Ottawa: International Reporting Inc, p.11

65 Kleist, Kuupik. "Welcoming speech to the Inuit Circumpolar Conference 11th General Assembly, Nuuk, Greenland, June 2010, available at: http://uk.nanoq.gl/Emner/News/News_from_Government/2010/06/Kuupik_speech_icc.aspx

66 Joint Review Panel for the Mackenzie Gas Project. 2006n. *Volume 16: Hearing held at Deline, NWT 3rd April 2006.* Ottawa: International Reporting Inc, p.1639.

67 In addition to the provisions within a number of conventions such as ILO Convention 169, there have been moves within the Arctic Council to consider the roles of existing indigenous land claim agreements as instruments to facilitate indigenous involvement and participation in planning and decision-making processes. For example, on the occasion of the Third Ministerial Meeting of the Arctic Council in Inari, Finland on 10 October 2002, the eight Arctic states adopted the Inari Declaration on Biodiversity Conservation and Sustainable Use of Natural Resources and agreed to:

> RECOGNIZE the potential for development of oil, gas, metals and minerals in many Arctic regions to impact on the standard of living and EMPHASIZE the importance of responsible management of these resources, including emergency prevention, to promote environmental protection and the sustainable development of the Arctic indigenous and local communities;

> CONSIDER the ecological and other impacts of natural resource development, and undertake, as appropriate, strategic assessments;

> ACKNOWLEDGE the need to pay particular attention to the impact of development and the use of natural resources on traditional sources of livelihood of indigenous peoples and their communities.

> (http://www.arctic-council.org/files/inari2002/inari_Declaration.pdf)

List of acronyms

AAC	Arctic Athabaskan Council
AEUB	Alberta Energy and Utilities Board
AFN	Alaska Federation of Natives
AGIA	Alaska Gasline Inducement Act
AHAPC	Alaska Highway Aboriginal Pipeline Coalition
AHGP	Alaska Highway Gas Pipeline
AHPP	Alaska Highway Pipeline Project
AIYA	Arctic Indigenous Youth Alliance
AMAP	Arctic Monitoring and Assessment Programme
ANCSA	Alaska Native Claims Settlement Act
ANGTS	Alaska Highway Gas Transportation System
ANILCA	Alaska National Interest Lands Conservation Act
ANWR	Arctic National Wildlife Refuge
AOGA	Alaska Oil and Gas Association
APG	Aboriginal Pipeline Group
ASRC	Arctic Slope Regional Corporation
AUC	Alberta Utilities Commission
BP	British Petroleum
Can$	Canadian Dollar
CARC	Canadian Arctic Resources Committee
CEAA	Canadian Environmental Assessment Agency
CNOOC	China National Offshore Oil Corporation
CO2	Carbon dioxide
COPE	Committee for Original Peoples' Entitlement
CPCN	Certificate of Public Convenience and Necessity
DEW	Distant Early Warning
DOI	U.S. Department of the Interior
EEZs	Exclusive Economic Zones
EIS	Environmental Impact Statement
ERCB	Energy Resources Conservation Board
GEM	Geo-mapping for Energy and Minerals
GSCI	Gwich'in Social and Cultural Institute
ICC	Inuit Circumpolar Council

ILO	International Labour Organisation
INAC	Indian and Northern Affairs Canada
IPCC	Intergovernmental Panel on Climate Change
ITK	Canada's national Inuit organization, Inuit Tapiriit Kanatami
IWGIA	International Work Group for Indigenous Affairs
JRP	Joint Review Panel
LNG	Liquid Natural Gas
MGP	Mackenzie Gas Project
MoU	Memorandum of Understanding
MPEG	Mackenzie Project Environmental Group
MVEIRB	Mackenzie Valley Environmental Impact Review Board
NANA	Northwest Alaska Native Association
NEB	Canada's National Energy Board
NEP	National Energy Program
NGO	Non Governmental Organisation
NGTL	Nova Gas Transmission Ltd.
NPA	Northern Pipeline Act
NPRA	National Petroleum Reserve-Alaska
NRBS	The Northern River Basins Study
NRC	National Research Council
NWT	Canada's Northwest Territories
PAS	Protected Areas Strategy
PSAs	Production Sharing Agreements
RAIPON	Russian Association of Indigenous Peoples of the North
SKDB	Sambaa K'e Dene Band
TAPS	Trans-Alaska Pipeline System
TEBTEBBA	Indigenous Peoples' International Centre for Policy Research and Education
UNCLOS	United Nations Convention on the Law of the Sea
UNDRIP	United Nations Declaration on the Rights of Indigenous Peoples
UNESCO	United Nations Educational, Scientific and Cultural Organization

USGS	United States Geological Survey
WWF	World Wildlife Fund
YESAA	Yukon Environmental and Socio-Economic Assessment Act
YNAO	Yamal-Nenets Autonomous Okrug

Bibliography

Abel, K. and K.S. Coates, 2001: Introduction: the North and the Nation. In *Northern Visions: new perspectives on the North in Canadian History,* ed. K. Abel and K. S. Coates. Peterborough: Broadview Press.

Abele, F., 1983: The Berger Inquiry and the politics of transformation in the Mackenzie Valley. Unpublished PhD thesis, York University.

ACIA, 2005: *Arctic Climate Impact Assessment: scientific report.* Cambridge: Cambridge University Press.

Adams, S., 1998: *Fort Chipewyan: a way of life study: summary report.* Vancouver: Stuart Adams & Associates.

AHDR, 2004: *Arctic Human Development Report.* Akureyri: Stefansson Arctic Institute.

Alaska Department of Natural Resources, 2006: *Division of Oil and Gas 2006 Annual Report.* Juneau: Alaska Department of Natural Resources.

Alaska Highway Aboriginal Pipeline Coalition, 2006: *Summary and comments on the Dena Tha' First Nation Decision.* Document # APC1.207: APC8.300, Whitehorse: Alaska Highway Aboriginal Pipeline Coalition.

Alberts, S., 2006: Oilsands crucial to U.S. plans. *Edmonton Journal,* 4 February, p. A3.

Anderson, M., M. Finer, D. Herriges, A. Miller and A. Soltani, 2009: *ConocoPhillips in the Peruvian Amazon.* A report by Amazon Watch and Save America's Forests, http://www.amazonwatch.org/conoco2009.pdf

Anisimov, O. and Fitzharris, 2001: Polar regions (Arctic and Antarctic). In *Climate Change 2001: Impacts, Adaptation, and Vulnerability. Contribution of Working Group II to the Third Assessment Report of the Intergovernmental Panel on Climate Change,* ed. McCarthy, J., O.F. Canziani, N.A. Leary, D.J. Dokken and F.S. White. Cambridge: Cambridge University Press.

AOGA, 1986: *Arctic National Wildlife Refuge.* Alaska Oil and Gas Association Advocacy Paper. Anchorage: AOGA.

Arctic Athabaskan Council, 2005: *Arctic Oil and Gas Assessment: social and economic effects of oil and gas activities in the Arctic.* Report and transcripts of a workshop held in Whitehorse, 23-24 June 2005. Whitehorse: Arctic Athabaskan Council.

Arctic Athabaskan Council, 2006: *Healthy, Sustainable Livelihoods for Indigenous Peoples: the impacts of global warming and oil and gas development.* Report of the Arctic Indigenous Leaders' K'atlodeeche Summit, Hay River, Northwest Territories, 11 and 12 December 2005. Whitehorse: Arctic Athabaskan Council. http://www.arcticathabaskancouncil.com/aacDocuments/public/Hay%20 River%20 Conference%20 Report.pdf

Beltaos, S., 2003: Numerical modeling of ice-jam flooding on the Peace-Athabasca Delta. *Hydrological Processes* 17: 3685-3702.

Berger, T., 1977: *Northern Frontier, Northern Homeland: the Report of the Mackenzie Valley Pipeline Inquiry.* 2 volumes, Ottawa: Department of Supply and Services.

Berger, T., 1989: The North as frontier and homeland in *The Arctic: choices for peace and security* [Proceedings of a public inquiry, held 18-19 March, Edmonton]. The

True North Strong and Free Inquiry Society. Vancouver and Seattle: Gordon Soules Book Publishers Ltd.

Berland, J., 2009: *North of Empire: essays on the cultural technologies of space*. Durham and London: Duke University Press.

Berton, P., 2001: *Klondike: the last great gold rush, 1896-1899*. Toronto: Anchor Canada.

Bird, Kenneth J., Charpentier, Ronald R., Gautier, Donald L., Houseknecht, David W., Klett, Timothy R., Pitman, Janet K., Moore, Thomas E., Schenk, Christopher J., Tennyson, Marilyn E. and Wandrey, Craig J., 2008: Circum-Arctic resource appraisal; estimates of undiscovered oil and gas north of the Arctic Circle. U.S. Geological Survey Fact Sheet 2008-3049, 4 p. [http://pubs.usgs.gov/fs/2008/3049/].

Bone, R., 2003: *The Geography of the Canadian North*. Oxford: Oxford University Press.

Bone, R., 2009: *The Canadian North: issues and challenges*. Oxford: Oxford University Press.

Boswell, R., 2009: Canada stands firm on northern frontier. *Edmonton Journal*, 24 November, p. A5.

Brody, H., 1981: *Maps and Dreams: Indians and the British Columbia Frontier*. Harmondsworth: Penguin.

Brody, H., 2000: *Assessing the Project – Social Impacts and Large Dams: Prepared for thematic review I.1: Social impacts of large dams equity and distributional issues*. Cape Town: World Commission on Dams, URL: http://www.dams.org/docs/kbase/contrib/soc192.pdf

Brooke, J., 2005: Big pipeline, major hurdles. *Edmonton Journal*, 20 February, p. D5.

Brower, C.D., 2002: Presentation made at the CERI North American Natural Gas Conference: Achieving Balance, & Calgary GasExpo, 4-5 March 2002,Telus Convention Centre, Calgary, Alberta.

Bunker, S.G. and P.S. Ciccantell, 2005: *Globalization and the Race for Resources*. Baltimore: Johns Hopkins University Press.

Byers, M., 2009: *Who Owns the Arctic? Understanding sovereignty disputes in the North*. Vancouver: Douglas and McIntyre.

Calliou, B., 2006: 1899 and the political economy of Canada's North-West: Treaty 8 as a compact to share and peacefully co-exist. In *Alberta Formed Alberta Transformed*, ed. M. Payne, D. Wetherell and C. Cavanaugh. Edmonton: University of Alberta Press.

Cameron, A.D., 1986[1909]: *The New North: an account of a woman's 1908 journey through Canada to the Arctic*. Lincoln and London: University of Nebraska Press.

Campbell, B.L., 1985: Uncertainty as symbolic action in disputes among experts. *Social Studies of Science* 15(3): 429-453.

Campbell, C.J., 2004: *The Coming Oil Crisis*. Brentwood, Essex: Multi-Science Publishing Co.

Campbell, C.J., 2005: *The Truth about Oil and the Looming Energy Crisis*. Skibbereen, Co. Cork: Eagle Print.

Campbell, P. C., 2000: *Footprints in the Delta* [Produced by J. Krepakevitch and K. Rankin]. Montreal: National Film Board, 44 min.

Card, J. R. and E. K. Yaremko, 1970: *Athabasca Delta Project: report #1*. Edmonton: Alberta Department of Agriculture, Water Resources Division.

Careless, J.M.H., 1954: Frontierism, metropolitanism, and Canadian history. *Canadian Historical Review* 35(1): 1-21.

Cargill, S., 2002: The Berger Inquiry Revisited: the meaning of inclusion for the Inuvialuit. Unpublished MA thesis, Dalhousie University.

Chalmers, J.W. (ed.), 1971: *On the Edge of the Shield: Fort Chipewyan and its hinterland.* Edmonton: Boreal Institute for Northern Studies.

Chambers, C., 1989: For our Children's Children: an educator's interpretation of Dene testimony to the Mackenzie Valley Pipeline Inquiry. Unpublished PhD thesis, University of Victoria.

Chapin, F.S., III, G. Peterson, F. Berkes, T.V. Callaghan, P. Angelstam, M. Apps, C. Beier, Y. Bergeron, A.-S. Crépin, K. Danell, T. Elmqvist, C. Folke, B. Forbes, N. Fresco, G. Juday, J. Niemela, A. Shvidenko, and G. Whiteman, 2004: Resilience and vulnerability of northern regions to social and environmental change. *Ambio* 33: 344-349.

Chapin, F. S. III, M. Berman, T. V. Callaghan, P. Convey, A. Crepin, K. Danell, H. Ducklow, B.Forbes, G. Kofinas, A. D. McGuire, M. Nuttall, R. Virginia, O.R. Young, and S. Zimov. 2005. Polar systems. in *Ecosystems and Human Well-Being: current state and trends*, ed. R. Hassan, R. Scholes and N. Ash. Millennium Ecosystem Assessment, Washington DC: Island Press.

Churchill, W., 2002: *Struggle for the Land: Native North American resistance to genocide, ecocide and colonization.*

Cleary, D., 1993: After the frontier: problems with political economy in the modern Brazilian Amazon. *Journal of Latin American Studies* 25: 331-49.

Coates, K.S., 1985: The Alaska Highway and the Indians of the southern Yukon, 1945-52: a study in native adaptation to northern development. In *The Alaska Highway: papers of the 40th anniversary symposium*, ed. K.S. Coates. Vancouver: University of British Columbia Press.

Coates, K.S. and W.R. Morrison, 1992: *The Forgotten North: a history of Canada's Provincial Norths.* Toronto: James Lorimer and Company.

Coates, K.S. and W.R. Morrison, 2005: *Land of the Midnight Sun: a history of the Yukon.* Montreal and Kingston: McGill-Queen's University Press.

Coates, K.S. and J. Powell, 1989: *The Modern North: people, politics and the rejection of colonialism.* Toronto: James Lorimer.

Coates, P.A., 1991: *The Trans-Alaska Pipeline Controversy: technology, conservation, and the frontier.* Fairbanks: University of Alaska Press.

Colt, S.G. and M. Pretes, 2005: Alaska Native Claims Settlement Act (ANCSA) in *Encyclopedia of the Arctic*, ed. M. Nuttall. New York and London: Routledge.

Commonwealth North, 1989: *An Alaskan View of ANWR.* Anchorage: Commonwealth North Inc.

Coulson, A. and R.J. Adamcyk, 1969: *The effects of the W.A.C. Bennett Dam on downstream levels and flows.* Ottawa: Canada Inland Waters Branch.

Council of Canadian Academies, 2009: *The Sustainable Management of Groundwater in Canada: Report of the Expert Panel on Groundwater.* Ottawa: Council of Canadian Academies.

Council for Yukon Indians, 1981: *Land before Money, Cooperation before Competition.* Whitehorse: Council of Yukon Indians.

Cowling, R., 2001: *Review and Regulatory Processes for Northern Pipeline Projects: opportunities for public involvement.* Calgary: Canadian Institute of Resources Law.

Crawford, T., 2010: Simplify environmental assessment: western premiers. *Edmonton Journal,* 17 June, p. A6.

Cuba, L.J., 1987: *Identity and Community on the Alaska Frontier.* Philadelphia: Temple University Press.

Dacks, G., 1981: *A Choice of Futures: politics in the Canadian North.* Toronto: Methuen.

De Angelis, M., 2004: Separating the doing and the deed: capital and the continuous nature of enclosures. *Historical Materialism* 12: 57-87.

Dey Nuttall, A. and M. Nuttall, 2009: Europe's Northern Dimension: policies, cooperation, frameworks. In *Canada's and Europe's Northern Dimensions,* ed. A. Dey Nuttall and M. Nuttall. Oulu: University of Oulu Press.

Earley, R., 2003: *DISCONNECT: Assessing and Managing the Social Effects of Development in the Athabasca Oil Sands.* Unpublished Master of Environmental Studies (MES) thesis, School of Planning, Faculty of Environmental Studies, University of Waterloo.

Ebner, D., 2005: Mackenzie fight is déjà vu all over again. *The Globe and Mail,* 12 May 2005, p. B5.

Espiritu, A., 1997: "Aboriginal Nations": Natives in northwest Siberia and northern Alberta. In *Contested Arctic: indigenous peoples, industrial states, and the circumpolar environment,* ed. E.A. Smith and J. McCarter. Seattle: University of Washington Press.

Finer, M. and M. Orta-Martínez, 2010: A second hydrocarbon boom threatens the Peruvian Amazon: trends, projections, and policy implications. *Environmental Research Letters* (5) 5 014012,10pp., http://www.iop.org/EJ/article/1748-9326/5/1/014012/erl10_1_014012.pdf?request-id=b7c3a3f5-380c-44c2-b193-d2ab31ca18e6

Fjellheim, R.S. and J. B. Henriksen, 2006: Oil and gas exploitation on Arctic indigenous peoples' territories: human rights, international law, and corporate social responsibility. *Journal of Indigenous Peoples' Rights* 4: 8-52.

Flyvbjerg, B., N. Bruzelius and W. Rothengatter, 2003: *Megaprojects and Risk: an anatomy of ambition.* Cambridge: Cambridge University Press.

Fold, N. and P. Hirsch, 2009. Re-thinking frontier in Southeast Asia. *The Geographical Journal* 175 (2): 95-97.

Fogarassy, T., 2007: *Triggering the duty of consultation with Aboriginal peoples.* Paper presented at the Canadian Institute's 10th BC Natural Gas Symposium 6-7 June, 2007, Vancouver, BC.

Fondahl, G. and A. Sirina, 2006: Oil pipeline development and indigenous rights in Eastern Siberia. *Indigenous Affairs* 1-2: 58-76.

Fondahl, G. and A. Sirina, 2006: Rights and risks: Evenki concerns regarding the proposed Eastern Siberia-Pacific Ocean Pipeline. *Sibirica* 5(2): 115-138.

Fontaine, G., 2005: Governance and the role of civil society: the case of oil and gas extraction in the Andean Amazon. In *The Handbook of Sustainability Research,* ed. W. Leal Filho. Frankfurt: Peter Lang Scientific Publishing.

Forbes, B.C., 1999: Land use and climate change in the Yamal-Nenets region of northwest Siberia: some ecological and socio-economic implications. *Polar Research* 18:1–7.

Forbes, B.C., 2004: Impacts of energy development in polar regions. In: *Encyclopedia of Energy,* ed. C.J. Cleveland. San Diego: Academic Press, 93-105.

Ford, J. and L. Berang-Ford, 2009: Food security in Igloolik, Nunavut: an exploratory study. *Polar Record* 45(234): 225-236.

Furniss, E., 1999: *The Burden of History: colonialism and the frontier myth in a rural Canadian community.* Vancouver: University of British Columbia Press.

Gatermann, R., 2010: Interview: Björn Tore Godal. *European Energy Review,* online edition, January 22, http://www.europeanenergyreview.eu/index.php?id=1656

Gautier, D.L., 2007: *Assessment of undiscovered oil and gas reserves of the East Greenland Rift Basins Province.* U.S. Geological Survey Fact Sheet 2007-3077, 4pp.

Gibson, R.B., 2002: From Wreck Cove to Voisey's Bay: the evolution of federal environmental assessment in Canada. *Impact Assessment and Project Appraisal* 20(3): 151-159.

Gilmartin, M., 2009: Border thinking: Rossport, Shell and political geographies of a gas pipeline. *Political Geography* 28(5): 274-282

Golovnev, A.V. and G. Osherenko, 1999: *Siberian Survival: the Nenets and their story.* Ithaca: Cornell University Press.

Gorman, M., 1987: Dene Community Development: Lessons from the Norman Wells Project. *Alternatives* 14(1):10-12.

Grace, S.E., 2001: *Canada and the Idea of North.* Montreal and Kingston: McGill-Queen's University Press.

Grant, G.M., 1873: *Ocean to Ocean: Sandford Fleming's Expedition through Canada in 1872.* Toronto: James Campbell and Son Ltd.

Grant, J., 2010: Mackenzie gas development needs responsible, sustainable plan. *Edmonton Journal,* 20 April, p. A12.

Gray, A., 1997: Onion Lake and the revitalization of Treaty Six. In *Honour Bound: Onion Lake and the Spirit of Treaty Six.* Copenhagen: IWGIA.

Griffith, B., D.C. Douglas, N.E. Walsh, D.D. Young, T.R. McCabe, D.E. Russell, R.G. White, R.D. Cameron, and K.R. Whitten, 2002: The Porcupine caribou herd. In *Arctic Refuge Coastal Plain Terrestrial Wildlife Research Summaries,* ed. D. C. Douglas, P. E. Reynolds, and E. B. Rhode. Biological Science Report USGS/BRD/BSR-2002-0001, U.S. Geological Survey, Biological Resources Division.

Gwich'in Steering Committee: www.gwichinsteeringcommittee.org

Gwich'in Steering Committee: 'Gwich'in Niintsyaa (Resolution)' http://www.gwichinsteeringcommittee.org/gwichinniintsyaa.html

Habermas, J. 1984: *The Theory of Communicative Action* vol. 1. London: Heinemann.

Habermas, J., 1990: *Moral Consciousness and Communicative Action.* Cambridge: Polity.

Hay, E. 2007. *Late Nights on Air.* Toronto: McClelland and Stewart.

Heitman Hansen, O. and M.R. Midtgard, 2008: Going North: the new petroleum province of Norway. In *Arctic Oil and Gas: sustainability at risk?* ed. A. Mikkelsen and O. Langhelle. London and New York: Routledge.

Holroyd, P. and H. Retzer, 2005: *A Peak into the Future: potential landscape impacts of gas development in northern Canada.* Drayton Valley, Alberta: The Pembina Institute.

Howard, R., 2009: *The Arctic Gold Rush: the new race for tomorrow's natural resources.* London: Continuum.

Howell, J.E. 1978: The Portage Mountain hydro-electric project. In *Northern transitions: volume I. Northern resource and land use policy study,* ed. E.B. Peterson and J.B. Wright. Ottawa: Canadian Arctic Resources Committee.

INAC, 2005: *Northern Oil and Gas Annual Report 2005.* Ottawa: Indian and Northern Affairs Canada.

Irlbacher-Fox, S., 2005: Land claims. In *Encyclopedia of the Arctic*, ed. M. Nuttall. New York and London: Routledge.

IWGIA, 1997: *Honour Bound: Onion Lake and the spirit of Treaty Six*. Copenhagen: IWGIA.

Jaremko, G., 2005a: Anxious NWT gov't urges NEB not to let Deh Cho kill pipeline. *Edmonton Journal*, 11 July, p. A16.

Jaremko, G., 2005b: North's demands too high. *Edmonton Journal*, 29 April, p. A3.

Jaremko, G., 2006: Finally, a way to get a grip on consultation. *Edmonton Journal*, 4 December, p. A16.

Jedrej, C. and M. Nuttall, 1996: *White Settlers: the impact of rural repopulation in Scotland*. Luxembourg: Harwood Academic.

Joint Review Panel for the Mackenzie Gas Project, 2006a: *Volume 2: Hearing held at Inuvik, NWT, 15th February 2006*. Ottawa: International Reporting Inc, p. 67

Joint Review Panel for the Mackenzie Gas Project, 2006b: *Volume 2: Hearing held at Inuvik, NWT, 15th February 2006* Ottawa: International Reporting Inc, p. 21-22

Joint Review Panel for the Mackenzie Gas Project, 2006c: *Volume 4: Hearing Held at Fort McPherson, NWT, 17th February 2006*. Ottawa: International Reporting Inc., p.283.

Joint Review Panel for the Mackenzie Gas Project, 2006d: *Volume 27: Hearing Held at Pehdzeh Ki Dene Band Complex, Wrigley NWT, 11th May 2006*. Ottawa: International Reporting Inc., p. 2461.

Joint Review Panel for the Mackenzie Gas Project, 2006e: *Volume 4: Hearing Held at Fort McPherson, NWT, 17th February 2006*. Ottawa: International Reporting Inc., p.288.

Joint Review Panel for the Mackenzie Gas Project, 2006f: *Volume 4: Hearing Held at Fort McPherson, NWT, 17th February 2006*. Ottawa: International Reporting Inc., p.291.

Joint Review Panel for the Mackenzie Gas Project, 2006g: *Volume 4: Hearing Held at Fort McPherson, NWT, 17th February 2006*. Ottawa: International Reporting Inc., p.301.

Joint Review Panel for the Mackenzie Gas Project, 2006h: *Volume 3: Hearing held at Inuvik, NWT, 16th February 2006* Ottawa: International Reporting Inc, p. 113

Joint Review Panel for the Mackenzie Gas Project, 2006i: *Volume 26: Held at Fort Simpson, NWT, 10th May 2006*. Ottawa: International Reporting Inc. p.2368.

Joint Review Panel for the Mackenzie Gas Project, 2006j: *Volume 51: Held at Ulukhaktok, NWT, 8th September 2006*. Ottawa: International Reporting Inc., p.4975.

Joint Review Panel for the Mackenzie Gas Project, 2006k: *Volume 51: Held at Ulukhaktok, NWT, 8th September 2006*. Ottawa: International Reporting Inc., p.4972.

Joint Review Panel for the Mackenzie Gas Project, 2006l: *Volume 51: Held at Ulukhaktok, NWT, 8th September 2006*. Ottawa: International Reporting Inc., p.4973.

Joint Review Panel for the Mackenzie Gas Project, 2006m: *Volume 1: Hearing held at Inuvik, NWT, 14th February 2006*. Ottawa: International Reporting Inc, p.11

Joint Review Panel for the Mackenzie Gas Project, 2006n: *Volume 16: Hearing held at Deline, NWT 3rd April 2006*. Ottawa: International Reporting Inc, p.1639.

Joint Review Panel for the Mackenzie Gas Project, 2009: *Foundation for a Sustainable Northern Future*. Ottawa: Government of Canada.

Jull, P., 1990: Inuit concerns and environmental assessment. In *The Challenge of Arctic Shipping: science, environmental assessment and human values*, ed. D. L.

Vanderzwaag and C. Lamson. Montreal and Kingston: McGill-Queen's University Press, pp. 139-153.

Kaalhauge Nielsen, J., 2005: Sustainable development. In *Encyclopedia of the Arctic,* ed. M. Nuttall. New York and London: Routledge.

Kakfwi, S., 2001: Maximizing Aboriginal benefits from resource development in the Northwest Territories. Presentation given at *First Nations Rights and Aboriginal Interests in Northern Pipeline Development Conference,* Calgary, Alberta, 18 June 2001.

Karl, T.L., 1997: *The Paradox of Plenty: oil booms and petro-states.* Berkeley: University of California Press.

Klare, M.T., 2004: *Blood and Oil: the dangers and consequences of America's growing dependency on imported oil.* New York: Henry Holt and Company.

Kleveman, L., 2003: *The New Great Game: blood and oil in Central Asia.* New York: Grove Press.

Kolausuk, E.D., 2003: Boom, bust and balance: life since 1950. In *Across Time and Tundra: the Inuvilauit of the Western Arctic,* ed. I. Alunik, E.D. Kolausuk and D. Morrison. Ottawa: Canadian Museum of Civilization.

Koivurova, T., 2009: Is there a race to resources in the polar regions? In *Canada's and Europe's Northern Dimensions,* ed. A. Dey Nuttall and M. Nuttall. Oulu: University of Oulu Press.

Kozij, J., 2009: Canada's Northern Strategy. In *Canada's and Europe's Northern Dimensions,* ed. A. Dey Nuttall and M. Nuttall. Oulu: University of Oulu Press, p.13.

Krupnik, I., 2000: Reindeer pastoralism in modern Siberia: Research and survival in the time of crash. *Polar Research* 19: 49-56.

Leas, D., 2005: Self-government in the Yukon. In *An Indigenous Parliament?: realities and perspectives in Russia and the circumpolar North,* ed. K. Wessendorf. Copenhagen: IWGIA.

Leonard, D.W., 1995: *Delayed Frontier: the Peace River Country to 1909.* Calgary: Detselig Enterprises Ltd.

Levy, D.A., 2009: *Pipelines and Salmon in Northern British Columbia: potential impacts.* Drayton Valley, Alberta: Pembina Institute.

Maas, D.C., 2005: Alaska National Interest Lands Conservation Act (ANILCA). In *Encyclopedia of the Arctic,* ed. M. Nuttall. New York and London: Routledge.

MacBeth, R.G., 1918: *The Romance of Western Canada.* Toronto: William Briggs.

Mackay, G.A., 1981: An economic perspective on migrant labour movements for the North Sea developments. In *Labour Migration and Oil,* ed. R. Moore. North Sea Oil Panel Occasional Paper No. 7, London: Social Science Research Council.

Mackenzie, A., 1911: *Voyages from Montreal through the Continent of North America to the Frozen and Pacific Oceans* 2 vols. Courier Press: Toronto.

Mair, C., 1908: *Through the Mackenzie Basin: a narrative of the Athabasca and Peace River Treaty Expedition of 1899.* Toronto: William Briggs.

McBeath, G. and T. Morehouse, 1994: *Alaska Politics and Government.* Lincoln: University of Nebraska Press.

McKenzie-Brown, P., 1998: *Richness of Discovery: Amoco's first fifty years in Canada.* Calgary: Amoco Canada Petroleum Co.

McQuade, L., 2004: *It's the Crude Dude: war, big oil, and the fight for the planet.* Toronto: Random House.

Marsden, W., 2007: *Stupid to the Last Drop: how Alberta is bringing environmental Armageddon to Canada (and doesn't seem to care).* Toronto: Alfred A. Knopf.

Mayeda, A., 2010: Arctic spill scenario stumps BP at hearing. *Edmonton Journal,* 14 May, p. A3.

Mikkelsen, A. and O. Langhelle (eds.), 2008: *Arctic Oil and Gas: sustainability at risk?* London and New York: Routledge.

Moe, A. and E. Wilson Rowe, 2009: Northern Offshore Oil and Gas Resources: Policy Challenges and Approaches. In *Russia and the North,* ed. E. Wilson Rowe. Ottawa: University of Ottawa Press.

Moore, R., 1981: *The Social Impact of Oil: the case of Peterhead.* London: Routledge and Kegan Paul.

Murashko, O., 2008: Protecting indigenous peoples' rights to their natural resources: the case of Russia. *Indigenous Affairs* 3-4/08: 48-59.

Nassichuk, W.W., 1987: Forty years of Northern non-renewable natural resource development. *Arctic* 40 (4): 274-84.

National Energy Board, 2002: *Cooperation Plan for the Environmental Impact Assessment and Regulatory Review of a Northern Gas Pipeline Project through the Northwest Territories. Northern Pipeline Environmental Impact Assessment and Regulatory Chairs' Committee, Calgary: National Energy Board.*

National Energy Board, 2006a: *Mackenzie Gas Project* Volume 1, Hearing held at Inuvik (Northwest Territories), Wednesday 25 January 2006. Ottawa: International Reporting Inc.

National Energy Board, 2006b: *Canada's Oil Sands—Opportunities and Challenges to 2015: an update.* Calgary: National Energy Board.

Nelkin, D., 1975: The political impact of technical expertise. *Social Studies of Science* (5): 37–54.

Nellemann, C.L., L. Kullerud, J. Vistnes, B.C. Forbes, and G.P. Kofinas (ed.), 2001: *GLOBIO—Global Methodology for Mapping Human Impacts on the Biosphere.* New York: United Nations Environment Programme.

Nikiforuk, A., 2008: *Tar Sands: dirty oil and the future of a continent.* Vancouver: Greystone.

Novikova, N.I., 1995: Okhotniki i netfianiki: vozmozhnostic dogovora. In *Sotsial'noekonomicheskoe i ku'lturnoe razvitie narodov Severa I Sibiri,* ed. Z.P. Solokova. Moscow: N.N. Mikluho-Maklaia RAN.

NRC, 2003: *Cumulative Environmental Effects of Oil and Gas Activities in Alaska's North Slope.* Washington DC: National Academies Press.

Nuttall, M., 1998: *Protecting the Arctic: indigenous peoples and cultural survival.* New York and London: Routledge.

Nuttall, M., 2006a: The Mackenzie Gas Project: Aboriginal interests, the environment and Northern Canada's energy frontier. *Indigenous Affairs* 2-3/06: 20-29.

Nuttall, M., 2006b: Alaska's Arctic National Wildlife Refuge Debate. *Indigenous Affairs* 3/06: 8-11.

Nuttall, M., 2008a: Aboriginal participation, consultation, and Canada's Mackenzie Gas Project. *Energy and Environment* 19 (5): 617-634.

Nuttall, M., 2008b: Self-Rule in Greenland: towards the world's first independent Inuit state? *Indigenous Affairs* 3&4/08: 64-70.

Nuttall, M., 2010: Epistemological conflicts and cooperation in the circumpolar North. In *Globalization and the Circumpolar North*, ed. L. Heininen and C. Southcott. Fairbanks: University of Alaska Press.

Nuttall, M., F. Berkes, B. Forbes, G. Kofinas, T. Vlassova and G. Wenzel, 2005: Hunting, Herding, Fishing and Gathering: indigenous peoples and renewable resource use in the Arctic. In ACIA *Arctic Climate Impact Assessment* Cambridge: Cambridge University Press.

Nuttall, M. and K. Wessendorf, 2006: *Arctic Oil and Gas Development.* Thematic Issue of *Indigenous Affairs* 2-3/06

Odell, P. 2010: Why we do not have to worry about 'peak oil'. *European Energy Review* Online edition, 15 January 2010, http://www.europeanenergyreview. eu/index.php?id_mailing=25&toegang=8e296a067a37563370ded05f5a3bf3ec& id=1627

Ott, R., 2005: *Sound Truth and Corporate Myths: the legacy of the Exxon Valdez oil Spill.* Cordova: Dragonfly Sisters Press.

O'Meara, D., 2010: Bitumen flow to the U.S. begins in 'historic' first for oilsands. *Edmonton Journal*, 1 July, p. E6.

Page, R., 1986: *Northern Development: the Canadian dilemma.* Toronto: McClelland.

Peace-Athabasca Delta Implementation Committee, 1987: *Peace-Athabasca Delta water management works evaluation: final report Peace-Athabasca Delta Implementation Committee.* Canada: Peace-Athabasca Delta Implementation Committee.

Piper, L., 2009: *The Industrial Transformation of Subarctic Canada.* Vancouver: University of British Columbia Press.

Polczer, S., 2010: India still in the hunt in oilsands.' *Edmonton Journal*, 10 June, p. E1.

Pollon, E.K. and S. Smith Matheson, 1989: *This Was Our Valley.* Calgary: Detselig Enterprises Ltd.

Povoroznyuk, O., 2006: Evenks of Chitinskaya Province: society and economy (still) in transition. *Indigenous Affairs* 1-2 /06: 68-74.

Pratt, L., 2001: *Energy: free trade and the price we paid.* Edmonton: The Parkland Institute.

Pretes, M., 1991: Northern Frontiers: political development and policy-making in Alaska and the Yukon. In *Borderlands: essays in Canadian-American relations*, ed. R. Lecker. Toronto: ECW Press.

Preston, J., 2009: Canada. In *The Indigenous World 2009*, ed. K. Wessendorf. Copenhagen: IWGIA.

Prowse, T. D., F.M. Conly, M. Church and M. C. English, 2002: A review of hydroecological results of the Northern River Basins Study, Canada. Part 1. Peace and Slave Rivers. *River Research and Applications* (18): 429-446.

Prowse, T., C. Furgal, C. Dickson, T. Edwards, L. Eerkes-Medrano, F. Jackson, H. Meling, D. Milburn, S. Nickels, M. Nuttall, A. Ogden, D. Peters, J. Reist, S. Smith, and M. Westlake, 2008: Northern Canada. In *From Impacts to Adaptation: Canada in a changing climate*, ed. D. Lemmen, F. Warren, E. Bush and J. Lacroix. Ottawa: Natural Resources Canada.

Rappaport, R., 1968: *Pigs for the Ancestors.* New Haven: Yale University Press.

Rasmussen, R.O., 2006: Oil exploration in Greenland. *Indigenous Affairs* 2-3/06: 40-47.

Roberts, P., 2004: *The End of Oil: on the edge of a perilous new world.* New York: Houghton Mifflin Company.

Roddick, D., 2006: Yukon First Nations and the Alaska Highway Gas Pipeline. *Indigenous Affairs 2-3/06*: 12-19.

Roon, T., 2006: Globalization of Sakhalin's oil industry: partnership or conflict? A reflection on the *Etnologicheskaia Ekspertiza. Sibirica* 5(2): 95-114.

Rohmer, R., 1973: *The Arctic Imperative: an overview of the energy crisis.* Toronto: McClelland and Stewart Limited.

Sale, R. and E. Potapov, 2010: *The Scramble for the Arctic: ownership, exploitation and conflict in the Far North.* London: Frances Lincoln Limited.

Schindler, D.W., 1998: Sustaining aquatic ecosystems in boreal regions. *Conservation Ecology* [online] 2(2): 18. Available from the Internet. URL: http://www.consecol.org/vol2/iss2/art18

Scott, P., 2007: *Stories Told: stories and images of the Berger Inquiry.* Yellowknife: The Edzo Institute.

Sirina, A., 2009: Oil and gas development in Russia and Northern indigenous peoples. In *Russia and the North*, ed. E. Wilson Rowe. Ottawa: University of Ottawa Press.

Stabler, J.C., 1977: The Report of the Mackenzie Valley Pipeline Inquiry, Volume 1: a socio-economic critique. *The Musk-Ox* 20: 57-65.

Stammler, F. and B.C. Forbes, 2006: Oil and gas development in the Russian Arctic: West Siberia and Timan-Pechora. *Indigenous Affairs* 2-3/06:48-57.

Stammler, F. and E. Wilson, 2006: Dialogue for development: an exploration of relations between oil and gas companies, communities and the state. *Sibirica* 5(2): 1-42.

Tesar, C., 2006: Arctic Leaders' Summit produces climate change declaration and action plan. http://www.arcticpeoples.org/2006/01/15/arctic-leaders%E2%80%99-summit-produces-climate-change-declaration-and-action-plan/, posted 15 January 2006.

Timoney, K., 2002: A dying delta? A case study of a wetland paradigm. *Wetlands* 22(2): 282-300.

Timoney, K., 2008: *Aboriginal Community Sustainability and Resource Extraction in Northern Alberta: Ecological and Landscape Component: A Report for the Community of Chipewyan Lake.* Sherwood Park, Alberta: Treeline Ecological Research.

Tuisku, T., 2003: Surviving in the oil age: co-existence of reindeer herding and petroleum development. In *Social and Environmental Impacts in the North*, ed. R.O. Rasmussen and N.E. Koroleva. Dortrecht: Kluwer Academic Publishers.

Tulk, D., 2005: Alaska and the Alberta Hub: delivering gas to market. *Alberta Oil and Gas* 1 (3): 33-35

Turner, F.J. 1920: *The Frontier in American History.* New York: Holt, Rinehart and Winston.

Tussing, A.R., 1984: Alaska's petroleum-based economy. In *Alaska Resources Development Issues: issues of the 1980s*, ed. T.A. Morehouse. Boulder: Westview Press.

U.S. Fish and Wildlife Service, 2001: Potential impacts of proposed oil and gas development on the Arctic Refuge's coastal plain: Historical overview and issues of concern. http://arctic.fws.gov/issues1.html

Usher, P., G. Duhaime and E. Searles, 2003: The household as an economic unit in Arctic Aboriginal communities, and its measurement by means of a comprehensive survey. *Social Indicators Research* 61: 175-202.

Venne, S. H., 1997: Introduction. In *Honour Bound: Onion Lake and the Spirit of Treaty Six.* Copenhagen: IWGIA.

Venne, S. H., 2006: Treaties made in good faith. In *Natives and Settlers Now and Then: historical issues and current perspectives on treaties and land claims in Canada,* ed. P. W. DePasquale. Edmonton: University of Alberta Press.

Villebrun, L.N., 2002: Flowing with the Land: the public transmission of Dene knowledge in environmental hearings. Unpublished Master of Arts thesis, University of Calgary.

Vistnes, I. and C. Nellemann, 2001: Avoidance of cabins, roads, and power lines by reindeer during calving. *Journal of Wildlife Management* 65:915–925.

Walker, A. 2006. Beyond Hills and Plains. *IIAS Newsletter* 42: 5.

Walker, D.A., P.J. Webber, E.F. Binnian, K.R. Everett, N.D. Lederer, E.A. Nordstrand and M.D. Walker, 1987: Cumulative impacts of oil fields on northern Alaskan landscapes. *Science* 238: 757-761.

Waterman, J., 2005: *Where Mountains are Nameless: passion and politics in the Arctic National Wildlife Refuge.* New York and London: W.W. Norton and Company.

WCED, 1987: *Our Common Future.* Oxford: Oxford University Press.

Webb, W.P., 1964: *The Great Frontier.* Austin: University of Texas Press.

Weber, B., 2006: Deh Cho reject deadline to join pipeline group. *Edmonton Journal,*Tuesday June 27, F2.

Weller, G., E. Bush, T. V.Callaghan, R. Corell, S. Fox, C. Furgal, A. H. Hoel, H. Huntington, E. Källén, V. Kattsov, D. Klein, H. Loeng, M. Long Martello, M. MacCracken, M. Nuttall, T. Prowse, L. O. Reiersen, J. Reist, A.Tanskanen, J. Walsh, B. Weatherhead and F. Wrona, 2005: Summary and Synthesis of the ACIA. In *ACIA Arctic Climate Impact Assessment: scientific report.* Cambridge: Cambridge University Press.

Westman, C., 2006: Assessing the impacts of oilsands development in indigenous peoples in Alberta. *Indigenous Affairs* 2-3/06: 30-39.

Wrona, F.J., Carey, J. Brownlee, B. and E. McCauley, 2000: Contaminant sources, distribution and fate in the Athabasca, Peace and Slave River Basins, Canada. *Journal of Aquatic Ecosystem Stress and Recovery* Vol. 8: 39-51.

WWF-Canada, 2010: Submission to the National Energy Board, 11 February 2010 https://www.neb-one.gc.ca/ll eng/livelink.exe/fetch/2000/90464/90550/33 8535/338661/588333/588334/594351/WWF-08_-_WWF Canada_Submission_ to_the_NEB__February_11_2010.pdf?nodeid=594807&vernum=0

Zaslow, M., 1948: The frontier hypothesis in recent historiography. *Canadian Historical Review* 29(2): 153-167.

Zaslow, M., 1971: *The Opening of the Canadian North, 1870-1914.* Toronto and Montreal: McClellan and Stewart.

Proposed Pipelines

Mackenzie Gas project

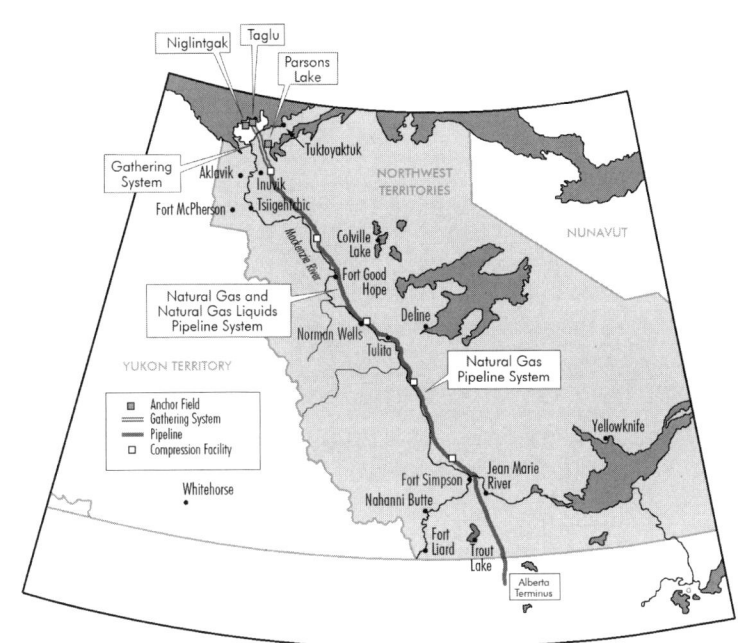

Arctic National Wildlife Refuge

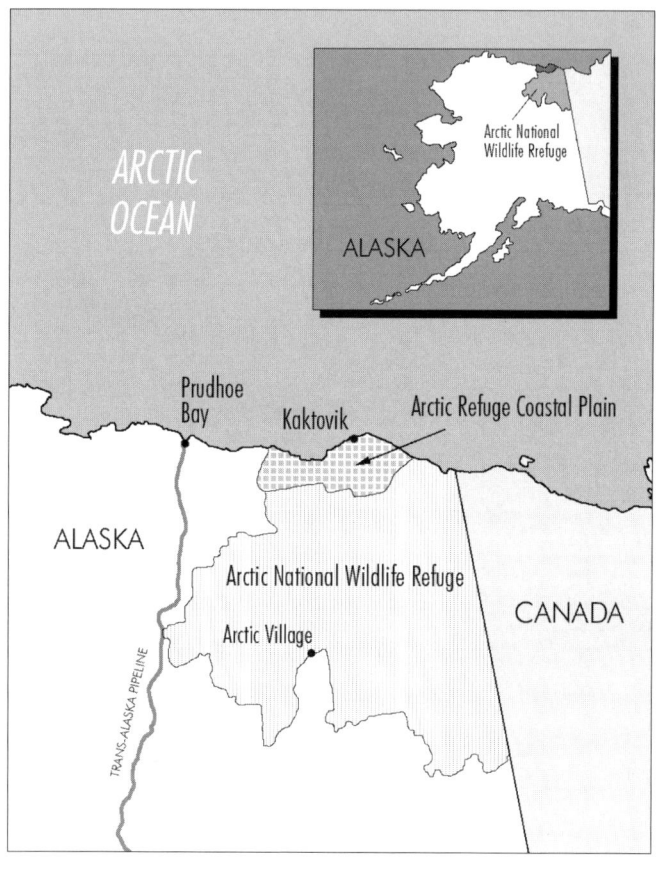

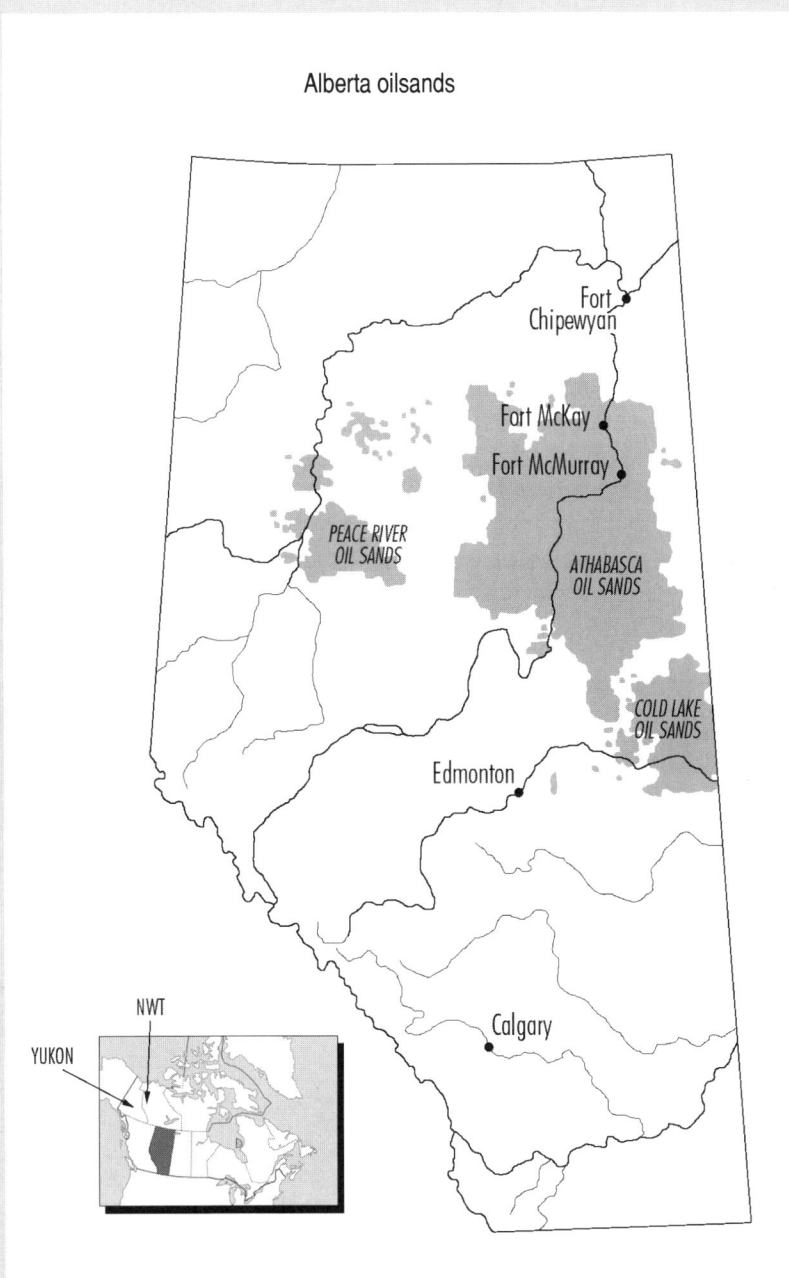

Alberta oilsands